Metals and Society: an Introduction to Economic Geology

Nicholas Arndt • Clément Ganino

Metals and Society: an Introduction to Economic Geology

 Springer

Nicholas Arndt
University of Grenoble
Grenoble
38031
France
arndt@ujf-grenoble.fr

Clément Ganino
University of Nice
Nice
06103
France
clement.ganino@unice.fr

Nicholas T. Arndt, Clément Ganino: Ressources minérales: Cours et exercices corrigés
Originally published: © Dunod, Paris, 2010

ISBN 978-3- 642-22995-4   e-ISBN 978-3- 642-22996-1
DOI 10.1007/978-3-642-22996-1
Springer Heidelberg Dordrecht London New York

Library of Congress Control Number: 2011942416

Printed on acid-free paper

Springer is part of Springer Science+Business Media (www.springer.com)

Preface

Thousands of years ago European's were transporting tin from Cornwall in southwest England to Crete in the eastern Mediterranean to create bronze by alloying tin and copper to create a new and more useful metal allow. Thousands of years from now humans we will still be using metals. The future will require existing metals for things we are used to having at our fingertips, pots and pans, vehicles and homes and also new types of uses of metals, some incorporated as nano-materials thus making them more effective as magnets for electric cars and wind and tide energy generation systems or as more malleable materials "plastic-metals".

The globalisation on the minerals industry is with us to stay, and supply and demand for raw materials will underlie economic, social and a political stability in much the same way as it did for the Minoans in the Bronze Age.

Geologists will be called upon to discover new mineral deposits and to think of new ways of mining minerals and remediation of the mining sites for which global pressures may require us to mine in pristine environments such as the deep sea-floor hydrothermal systems, in the Arctic, or even the Antarctic. We will use novel extraction technologies through robotics, in-situ leaching, or concentration from dilute natural systems such as sea-water.

It is thus essential that research in ore deposits (economic geology) is maintained in earth science departments across the globe and that scientists have an appreciation for the natural process of concentration of metals and the economics of the resource in order to maintain active exploration and mining programmes. This involves understanding the need for, and trade in, the resource and also the tectonic, volcanic and sedimentary processes that concentrate metals to make an ore that is of high enough grade to be economically feasible to extract.

This book provides an excellent overview of the subject for the general geologist. It includes some thought-provoking statements and questions for discussion on globalisation and the current practices of the minerals industry.

Nottingham, UK   John Ludden

Contents

Introduction

In the years that preceded the writing of this book, metal prices first soared to record levels, then plummeted to half these values (Fig. 1.1). Accelerating demand from China and other developing countries triggered the rise; collapse of the world economy triggered the fall. When prices were high, mineral exploration companies doubled their efforts to find new resources, and geologists were in great demand; the fall has stifled this demand. As the economy gradually recovers, driven by the rapid growth of the Chinese economy, new deposits are again sought, and there is once again a need for trained geologists. Most earth science students have a broad geological education that includes high-level courses in the subjects required of an exploration geologist – structural geology, field mapping, remote sensing, geophysics. What is missing is an elementary knowledge of economic geology.

We wrote this book to fill a gap in the literature available to students of the earth sciences. Many excellent and modern books describe in detail the characteristics of ore deposits and others discuss modern theories on how the deposits might have formed. Some books deal briefly with the economic issues that govern the mining of ores and the mineral industry in general, but usually this treatment is secondary.

As we explain in the first chapter, the very definition of an ore and of an ore deposit is grounded in economics – an ore is natural material that can be mined at a profit. Any comprehensive treatment of the subject must include discussion of what distinguishes an ore deposit from any other body of rock, a discussion that includes not only the geological aspects but also the geographic, economic and financial elements that influence the viability of a mining operation. To be able to follow such a discussion requires at least a basic knowledge of the commercial aspects of mining operations and of world trade in mineral products. Our aim in this book is to provide basic information about the scientific issues related to the nature and origin of ore deposits, to explain how, where and why metals and mineral products are used in our modern society, and to illustrate the extent to which society cannot function without these products.

The expansion of exploration and development of ore deposits will coincide with an increasing awareness of the fragility of our planet's environment, particularly the threat posed by global warming. Calls for "sustainable development" will accompany this economic revival, and the mining, transport, refining and consumption of raw materials will be subject to close scrutiny. At present most university students are taught almost nothing of this issue (or if they are taught, in courses on ecology and the environment, the reference to mining is totally and massively negative). The exploitation of ore deposits in the past has caused great damage to small parts of the Earth's surface, and mining with no regard to the environment can no longer be permitted. But if the world requires steel, aluminium or rare earths – to build wind turbines, for example – or copper and silica to build solar panels, the raw materials must be mined. These and other issues are discussed in our book.

Throughout the book, exercises are provided to illustrate the complexities, contradictions and dilemmas posed by society's needs for natural resources. We discuss the issue of when, or more exactly if ever, our supplies of metals will be exhausted. We consider the notion of sustainable development and the environmental damage done by many mining operations. At present the needs of the industrialized "first-world" countries are met in large part by the importation of ores from lesser-developed countries; we consider the economics and the ethics of this trade. The first author is an unabashed free-marketer; the views of the second, French, author are more nuanced. Throughout the book we have not hesitated to express our views. To a student who has received all his or her knowledge of mineral economics and global trade from local media and other popular sources of information, many of these views will come as a surprise, even as a shock, but we have not toned down the our treatment to conform to prevailing viewpoints. Instead we have written many relevant sections in a deliberately provocative manner in order to encourage discussion of these important issues.

In the first two chapters and in the last, geological and economical issues receive equal billing. In these chapters we define ores and ore deposits, discuss how they are classified, and explain that the study of ore deposits is intrinsically linked with the global economy. We explain how the viability of an ore deposit depends directly on the metal price, which in turn is linked to the demand from society for the mineral product. The factors that control this demand and the way the demand is satisfied by the discovery of new mineral deposits is a major subject in these chapters. Chapter 2 is an overview of the global distribution of ore deposits – where they are mined, where they are refined, and where the final products are consumed.

The following three chapters are more geological. In them we discuss the nature and origin of three broad groups of ore deposits: those that form through magmatic processes, those that result from the precipitation of minerals from hydrothermal fluids, and those that form in a sedimentary or superficial environment. The emphasis is on the ore-forming process and exhaustive descriptions of the ore deposits themselves are largely missing. We also chose not to include abundant references to published papers but instead provide a selection of important sources in information at the end of each chapter. The principles of ore-forming processes are illustrating by way of discussion of a selection of well-known examples.

In the final chapter, which deals with the future of economic geology, we consider two 'new' types of strategic ores – rare earth elements and lithium – that will become increasingly important for the electronics and transport industries of the twenty-first century. We chose these examples because they illustrate well the paradoxes and challenges posed by the need to supply society with strategic materials at a time when the global balance of power is rapidly changing.

We thank Chris Arndt, Anne-Marie Boullier, Marie Dubernet, Mélina Ganino, Jon Hronsky, Emilie Janots, Elaine Knuth, John Ludden, Jérôme Nomade, Michel Piboule, Gleb Pokrovski and Chystele Verati for their carefully reading the first version of this book and for their useful comments and suggestions. We also thank Grant Cawthorn, Axel Hofmann, Kurt Konhauser, Phil Crabbe and Peter Mueller for the photographs they provided. The French Centre Nationale de Recherche Scientifique (CNRS), the Université Joseph Fourier in Grenoble and the Université de Nice – Sophia Antipolis supported us during the preparation of the manuscript.

Chapter 1
Introduction

1.1 What Is Economic Geology?

We start this chapter with Fig. 1.1, which shows how the price, the average grade and production of copper ore changed from 1900 to the present. At the start of last century the price was about $7,000 per ton (expressed in today's currency); by 2002 it had decreased threefold to about $1,800 per ton, then, in the past 3 years to 2010 (when this book was written), it rose sharply to about $9,000 per ton. Over the same period, the total amount of copper mined gradually increased, except in the early 1920s and 1930s when both price and production dropped. Figure 1.2 shows that other metals followed similar trends. How do we explain these changes, and what do they tell us about how the metal is found and mined, and about how it is used by society? Understanding these concepts is the basis of economic geology.

To explain these trends – the broad correlation between price and grade, the anti-correlation between price and production, and the periods that bucked the trend in the 1930s and in the past few years – we first consider the declining prices. Why was the price of copper in the year 2000 only 30% of the price at the start of the previous century? The more important, and apparently contradictory elements in the explanation are:

- *Exhaustion of rich and easily mined deposits.* As these deposits are mined out, we have turned to deposits with lower concentrations of copper. The average grade has decreased from about 1% at the turn of the nineteenth century to about 0.7% or less at the start of the twenty-first century. At the same time, most deposits close the centres of industry in Europe or American have been exhausted and new mines have opened far from the regions where the metal is used, often in regions with hostile climate or difficult mining conditions. Normally one would think that these trends would be associated with increasing scarcity of copper – a decrease in supply that should, according to the economic rule of supply and demand, have led to a price increase. Yet, from the start of the century, the opposite has happened. Why?

N. Arndt and C. Ganino, *Metals and Society: an Introduction to Economic Geology*,
DOI 10.1007/978-3-642-22996-1_1, © Springer-Verlag Berlin Heidelberg 2012

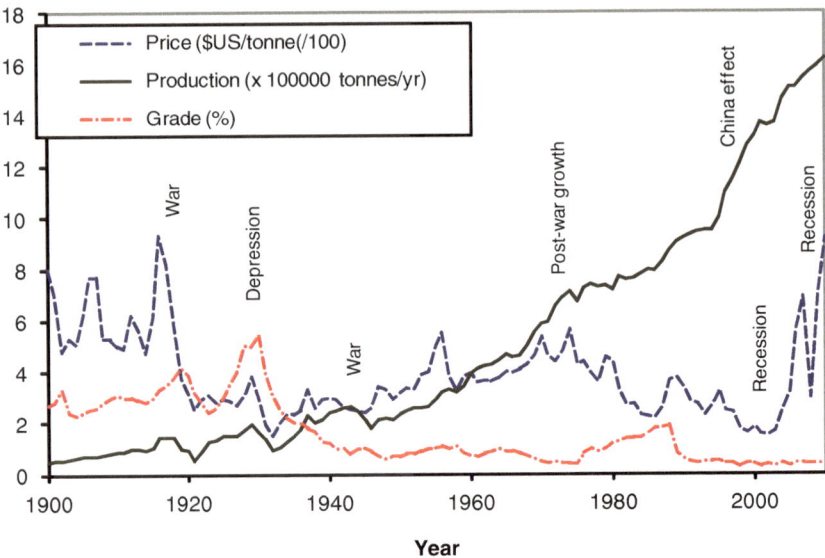

Fig. 1.1 Evolution in the price and production of copper over the past 120 years (statistics from the United States Geological Survey 2010, Mineral Resources Program. http://minerals.usgs.gov/products/index.html)

- *Improvements in technology.* The main reason why the price of copper has dropped steadily is improvement in the efficiency of the mining and refining industry, a chain of operations that starts with the search for new deposits, continues through the mining of these deposits and ends with the extraction of the metal from the mined ore. At the turn of the last century it was only possible to mine deposits with high grades that were close to the surface and close to industrial centres. Exceptions were a few unusually large and unusually rich deposits in more remote areas. Improvements in mining and extraction technologies have changed all this. Today's copper mines are enormous operations – vast open-pits that extract hundreds of thousands of tons of ore per day. Through the advantages of scale and the utilisation of modern techniques, it is possible now to mine ore with as little as 0.5% Cu. And with the economy of scale and improvement of technology has come a decrease in the cost of mining, an increase in supply, and a century-long drop in the price of the metal.

Now let us consider in detail the trends illustrated in Fig. 1.1. The decrease in copper price in the 1930s, and the corresponding decrease in copper production coincided with the Great Depression. Economies throughout the world collapsed, demand for copper plummeted and this had immediate repercussions on the price. The opposite has happened in the past 5 years. The economic miracles in China and to a lesser extent in India have boosted the industrial and societal demands of two billion people. To construct the cell phones, cooking pans and power stations that

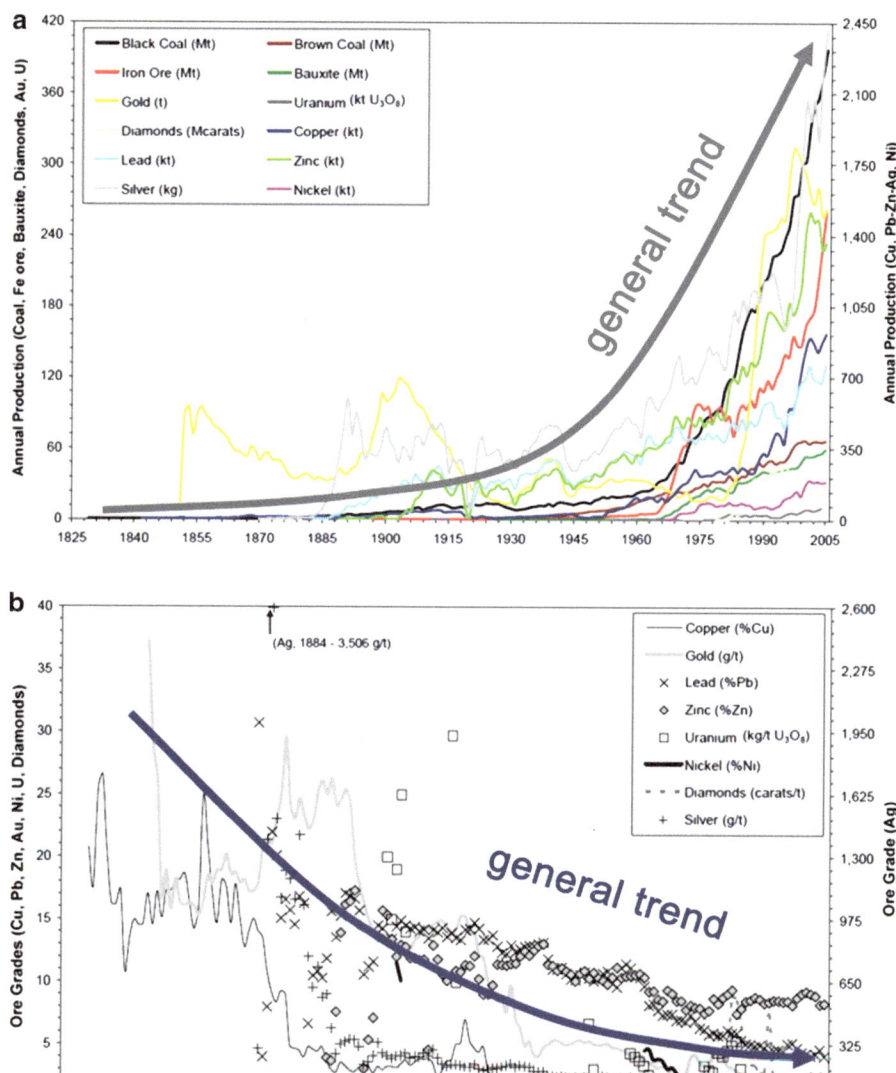

Fig. 1.2 (**a**) Evolution of production of selected metals since the mid-nineteenth century, (**b**) evolution of ore grades for the same metals (Modified from Mudd 2010)

they now expect (so as to live in more or less the same way as people in Europe and America) requires a vast acceleration in the rate at which copper is mined. Demand has exploded and this has triggered an immediate increase in the price of the metal.

How has this demand been met? New deposits of copper cannot be found overnight. The average time between the inauguration of a new exploration program and the start of mining of a new deposit is 10–15 years. Copper production has increased steadily over the past two decades, initially during a period of falling prices, and more recently during a period when the price of copper has tripled. In the first period, exploitation of stockpiles, the introduction of new improved mining and extraction techniques, and the opening of new large high-production mines, particularly in South America and Oceania, made this possible. Throughout the 1990s many mines were running at a loss: the cost of production was greater than the value of the metal extracted from the mine. Then from 2005 onwards, as the copper price increased, mines that had been loss-making operations suddenly started making money. Improvements in technology, which made it possible to mine and refine the ore more efficiently, aiding the return to profitability. Other deposits that had been explored and evaluated by mineral exploration companies but had been put aside because they were not viable at low copper prices suddenly became viable. Nothing had happened to the deposit: it still contained the same grade of copper and the same total amount of copper, and its location both geographically and geologically also had not changed. But a deposit that in the year 1998 was of little economic interest had became potentially highly profitable in 2010. These ideas lead us to examine several notions and definitions that are fundamental to economic geology.

Box 1.1 Consider the Following Statements and Discuss What They Tell Us About Economic Geology and the Mining Industry, as Perceived by the General Public

1. In the 1990s a Japanese scientist developed a new type of catalytic converter in which manganese replaced platinum. Why is this discovery important?
2. English ecologists have proposed that a new tax should be applied to "rare" metals such as silver, lead and copper. What do you think of this suggestion?
3. A journalist recently suggested that war might break out over the last drops of petrol. Is this suggestion reasonable and realistic?

Response

Consider the first statement. Why would it be important if manganese could be used in the place of platinum in the catalytic converters that are fitted to every new car? The answer lies in the price of the two metals. In February 2008, platinum (Pt) sold for about €100 per gram and manganese (Mn) for 10 cents per gram (€10,000 per ton), a 1000-fold difference in price. If Mn could replace Pt, catalytic converters would be much cheaper. Currently the cost of the metal makes up about half the cost of the converter, so if Mn replaced Pt, the cost would be cut by almost half. (Unfortunately the process does not work and Pt continues to be a highly sought-after metal). This discussion

(continued)

leads to the following question: why is platinum so much more expensive that manganese?

Consider now the other two statements. Both focus on the idea that resources of natural products such as metals and petroleum will soon be totally mined out or exhausted. "Peak oil", the notion that global production of petroleum has already, or very soon will, pass through a maximum, expresses the same idea. (You may have seen a TV program showing a sad fleet of aircraft stranded at an airport, the last drops of kerosene having been used up). Is this idea reasonable?

In the following section we discuss the notion that supplies of various types of natural resources will be depleted or exhausted in the near future. We conclude that none of the metals mentioned by the ecologists should be described as "rare" and that petroleum supplies will never be completely exhausted.

1.2   Peak Copper and Related Issues

One of the few natural products that went through a peak of production then dramatically declined is, paradoxically, renewable. Spermaceti, a wax present in the head cavities of the sperm whale, was an important product of the whaling industry throughout the eighteenth and nineteenth centuries. It was valued as high-quality lamp oil and later used as a lubricant. "Peak spermaceti" occurred at the start of the twentieth century when overfishing drastically reduced the number of sperm whales. The price rose drastically and this led to a search for substitutes; electric lighting replaced oil lamps, and oil from the jojoba plant was used as a lubricant. The demand for the product diminished, in part a consequence of social pressure to ban or restrict whaling. Now, as stocks of sperm whale slowly rebuild, not even Japanese whalers talk of hunting them.

Box 1.2 Peak Spermaceti and Peak Oil

We have drawn a comparison between the production and consumption of two very different products, petroleum and spermaceti. One is a natural product, essentially renewable (if sperm whales are not hunted to extinction). The other is a fossil resource that required millions of years to develop and is no longer being produced in any quantity. One is a product that was used widely in the nineteenth century, but only by a small and privileged part of the world's population. The other is currently used throughout the world. It is consumed by people rich and poor and is essential for our modern industrialized society. The exhaustion of petroleum resources, if this were ever to happen, would have a far more drastic impact than an absence of spermaceti.

Is it ridiculous to associate spermaceti and petroleum (as suggested by one reviewer of the book), or does the comparison have some merit? Discuss.

A parallel can be made with the exploitation of any natural product, including metallic ores as well as petroleum. Although there can be little doubt that the production of oil and gas will eventually pass through a peak, maybe this decade, maybe far later, it is by no means clear that the cause of the peak will be the exhaustion of petroleum resources. As supply diminishes, or is perceived to diminish, price will increase and this will inevitably, sooner or later, lead to a drop in demand. Use of petroleum will decline as we learn to waste less energy or find alternative energy sources; and, in much the same way as pressure from public and scientific bodies led to the banning of sperm whaling, pressure from the same groups will lead us to limit petroleum use so as to decrease the rate of global warming.

Another parallel can be drawn with slate, which in past centuries was widely used as roofing material. No one would argue that "peak slate" in the early twentieth century was due to exhaustion of the resource. The cost and effort of constructing slate roofs simply became prohibitive and alternative roofing materials were developed. Or, to use another commonly cited example, the Stone Age did not end for lack of stone.

The notion that we will run out of natural resources, including metals, is not new. Malthus (1830) in his celebrated article written in 1798 (Malthus, 1930; *An Essay on the Principle of Population, as it Affects the Future Improvement of Society with Remarks on the Speculations of Mr. Godwin, M. Condorcet, and Other Writers*) predicted that the increase in human population would rapidly exhaust supplies of food and natural resources, and the theme has been repeated many times since then. In the report of the 'Club of Rome', published as the book "Limits to Growth", Meadows et al. (1972) used a model in which human population and consumption of resources increased exponentially while the rate of discovery of new resources increased linearly or not at all. The consequence, if these assumptions are correct, is the rapid exhaustion of these resources, as shown in Fig. 1.3. According to the prediction made in 1970, the year that the book was written, global supplies of copper would now be nearly exhausted. Clearly this has not happened – copper is still mined in deposits all over the world. In 1970, the total amount of copper known to exist in clearly identified and readily exploitable deposits was sufficient to assure supplies, at the rate of consumption estimated at that time, for only the following 21–48 years, depending on which assumptions are made. Table 1.1 compares the predicted times before exhaustion of copper and six other metals, as estimated by Meadows et al. (1972), with another set of estimates made in 2009 by Mining Environmental Management, an industry journal. Despite almost 40 years of increasing consumption, the estimated times before exhaustion of these metals have barely changed and in some cases they have increased. How can this be?

Several factors have pushed back the supposed date of copper exhaustion. First and foremost, new copper deposits have been found and developed at such a rate that the predicted exhaustion time of known resources has remained constant. It must be recognized that it makes absolutely no sense for a mineral company or government agency to spend money to find resources that will not be exploited in the relatively near future. Once a company, or a government agency, has found

Fig. 1.3 (**a**) The predictions of Meadows et al. (1972) of the evolution of global population and of the supplies of raw materials. (**b**) Predictions based on the idea that supplies of natural resources will be rapidly exhausted, leading to a catastrophic decline in population

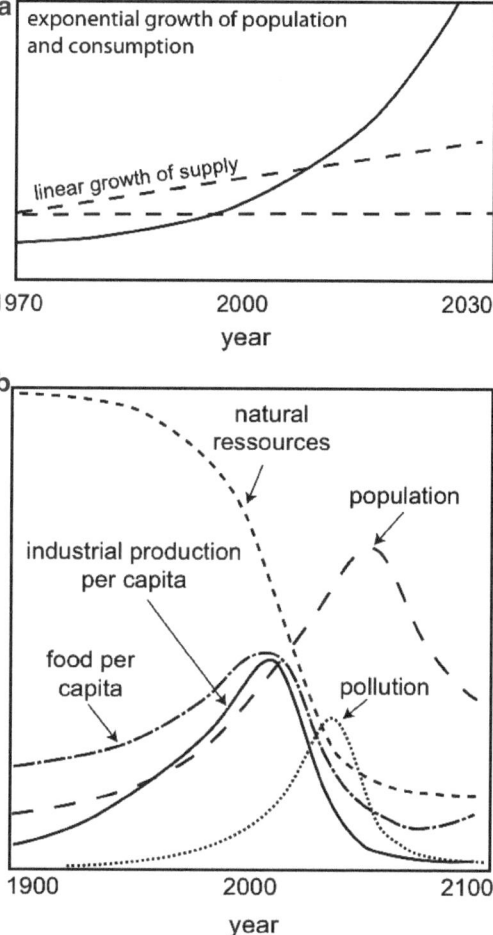

sufficient copper for the next two to three decades in deposits that can be exploited using current technology, there is no point in finding more.

The second influence that was not sufficiently well taken into account by Meadows and co-authors is the impact of improvements in technology, which has allowed even low-grade deposits to be mined efficiently, and the metals and other mineral products to be extracted economically. Later chapters provide striking examples of the evolution of mining and extraction technologies.

A fundamental difference between the long-term production of metals and energy sources such as petroleum, coal or uranium, is that once an energy source has been used by industry or society, it is gone for good. The fossil fuels disappear up smokestacks as they produce heat; the radioactive elements decay definitively to their daughter products. Metals, on the other hand, persist. Copper remains copper when it is used in telephone wires, in iPhones or on cathedral roofs, and in most cases it can be recovered at the end of the product's lifetime. The proportion of

Table 1.1 Time before exhaustion of a selection of metals, as estimated in 1972 and 2009

	Meadows et al. (1972)				Mining Environmental Management	
	Number of years (1972 – S)	Year when metal is exhausted (S)	Number of years (1972 – L)	Year when metal is exhausted (L)	Number of years (2009)	Year when metal is exhausted
Aluminium	31*	2003+	55	2027	131	2140
Copper	21	1993	48	2020	32	2041
Gold	9	1981	29	2001	16	2025
Iron	93	2065	173	2145	178	2187
Nickel	53	2025	96	2068	41	2050
Silver	13	1985	42	2014	13	2022
Zinc	18	1990	50	2022	17	2026

*Number of years before the metal becomes expensive and its supply limited
1972 (S) – exponential index of Meadows et al. (1972)
+Year (S) – year during which metal is exhausted
1972 (L) – exponential index of Meadows et al (1972) using an estimate of resources five times greater than those known in 1972
2009 – estimate of Mining Environment Management

copper and other metals that is recycled and reused by industry will continue to mount in future decades.

Many authorities now predict that supplies of metals and other mineral products are sufficient to meet societal needs for the foreseeable future. Other negative consequences of population increase, correctly identified by Meadows et al. (1972), will be far more drastic. Even though the rate of population increase will diminish with improvement in the standard of living and level of education in developing countries, the addition of one to three billion people will put a severe strain on all the earth's resources. Increasing competition for water and food, the increasing effects of pollution, climate change, the increased energy requirements for processing low-grade ores, and to a far lesser extent an increasing scarcity of petroleum, will severely test humanity's capacity to adapt. Nonetheless, although the long-term outlook is difficult to predict, we argue that the supplies of copper and most other mineral products will NEVER be totally exhausted. To understand this argument we must now consider in more detail the nature of an ore deposit.

1.3   What Is an Ore?

According to one commonly accepted definition, an ore is *a naturally occurring solid material containing a useful commodity that can be extracted at a profit.* There are several key phrases in this definition. By "useful commodity" we mean any substance that is useful or essential to society, such as metals, or energy sources, or minerals with distinctive properties.

Table 1.2 Properties and uses of a selection of substances (elements and minerals)

Type	Useful substance	Uses and properties
Alkali metals	Cesium (Cs)	Radioactive source (atomic clocks, medicine)
	Lithium (Li)	Batteries
	Potassium (K)	Pharmaceutical Industry
	Rubidium (Rb)	Photovoltaic cells, safety glass
	Sodium (Na)	Pharmaceuticals, cosmetics, pesticides
Alkali earths	Barium (Ba)	Trapping of residual gases in cathode ray tubes
	Beryllium (Be)	Alloys
	Calcium (Ca)	Alloys
	Magnesium (Mg)	Chemical and pharmaceutical industries, light alloys
	Radium (Ra)	Luminescence (watches)
	Strontium (Sr)	Varnishes, ceramic glazes
Base metals	Cadmium (Cd)	Batteries, alloys
	Cobalt (Co)	Alloys, catalyst in the chemical and petroleum industry
	Copper (Cu)	Electrical conductors, alloys
	Lead (Pb)	Car batteries, plumbing[a], crystal (glass), ammunition[a]
	Molybdenum (Mo)	Alloy (hardened steel), catalyst (oil industry)
	Nickel (Ni)	Alloys (stainless steel), batteries, electric guitar strings
	Tin (Sn)	Bronze (copper and tin), coating of tin cans[a], electronics (solder), coins
	Zinc (Zn)	Galvanizing (protection of steel against corrosion by depositing a thin layer of Zn), brass (copper-zinc alloy)
Construction metals	Iron (Fe)	construction – cars, buildings, bridges
	Aluminium	aircraft, electric cables
	Chromium (Cr)	Alloy (stainless steel), protective coating on steel
	Manganese (Mn)	Alloys, batteries, fertilizer
	Vanadium (V)	Additive in steel, catalyst
Other metals	Bismuth (Bi)	Fuses, glass, ceramics, pharmaceutical and cosmetic industries
	Hafnium (Hf)	Filament in light bulbs, nuclear reactors, alloys, processors
	Mercury (Hg)	Pharmaceutical industry, cathode fluorescent lamps, dental fillings[a], batteries, thermometers[a]
	Niobium (Nb)	Alloys, superconducting magnets
	Scandium (Sc)	Alloys (especially aluminum), metal halide lamp
	Tantalum (Ta)	Electronic capacitors
	Technetium (Tc)	Medical Imaging
	Thallium (Tl)	Low temperature thermometers, infrared detectors
	Titanium (Ti)	Pigments, high-technology alloys
	Tungsten (W)	Tungsten carbide – abrasive
	Yttrium (Y)	TV screens, lasers (YAG), superconducting alloys
	Zirconium (Zr)	High-technology alloys
Precious metals	Gold (Au)	Jewelry, coins, gold
	Indium (In)	Photovoltaic cells, infrared detectors, nuclear medicine
	Iridium (Ir)	Alloys (hardening of platinum alloys), mirror finish on ski goggles
	Osmium (Os)	Alloys, pen nibs, pacemakers
	Palladium (Pd)	Electronics (cell phones, computers …), catalysts, hydrogen sensors, jewelry

(continued)

Table 1.2 (continued)

Type	Useful substance	Uses and properties
	Platinum (Pt)	Electronics (cell phones, computers …), catalysts, hydrogen sensors, jewelry
	Rhenium (Re)	Alloys
	Rhodium (Rh)	Catalysts, X-ray tubes, mirrors, jewelry
	Ruthenium (Ru)	Alloys, hard drives, superconductors
	Silver (Ag)	Jewelry, silverware, photography[a]
Minerals	Diamond	Jewelry, abrasives (hardness, attractiveness)
	Corundum	Abrasives (hardness)
	Talc	Lubricant (softness)
	Pumice	Abrasives (hardness)
	Asbestos	Insulator (low thermal conductivity)[a]
	Mica	Insulator (low thermal conductivity)[a]
	Diatomite	Filters
	Barite	Drilling mud (high density)
	Andalusite	Ceramics (resistance to high temperature)
	Kyanite	Ceramics (resistance to high temperature)
	Halite	Food additive, de-icer (lowers freezing temperature of water)
	Calcite	Cement

[a]Use now restricted because of toxicity of substance or substitution

The uses of copper are well known. Without this metal (or other metals with similar properties) there would be no television sets, power stations and airliners, not to mention brass cooking pots and green-coloured domes on old cathedrals. Other metals such as iron, manganese, titanium and gold find a multitude of applications in the world in which we live. Some of these are listed in Table 1.2 and in an excellent web sites of the United States Geological Survey http://minerals. usgs.gov/granted.html and the British Geological Survey 2010; http://www.bgs.ac. uk/mineralsuk/statistics/worldStatistics.html. Ores also include energy sources, specifically coal and uranium. Petroleum is normally excluded from the definition, which is generally restricted to solids, but the bitumen recovered in deposits such as the Athabasca tar sands might be considered an ore. Finally there is a range of products, generally of low cost, that are also mined and also constitute ores: included in this list are building materials such as limestone for cement, gravel for road construction and the building industry, ornamental stones and gems, fertilisers, abrasives, even common salt.

Box 1.3 The Criticality Index of the United States Geological Survey
A committee of geologists and economists from various governmental agencies and universities in the USA published a report evaluating the supply situation of a wide range of metals and mineral products (National Research Council 2008; http://dels.nas.edu/dels/rpt_briefs/critical_minerals_final. pdf.). Although the report focussed specifically on the US situation, the broad conclusions apply also to European countries. The committee defined
(continued)

the "criticality index" which is the product of the importance of the product in an industrial society (the x-axis) and the degree to which its supply is subject to potential restrictions (the y axis). The importance depends not so much on the amount that is used but more on whether the product is used in critical applications and whether it can be substituted by other materials. The supply risk depends on factors such as whether the product is produced locally or must be imported, the geographic location of sources, and the political stability of the supplying country or region. In the graph below, we see that copper is relatively important but is subject to little supply risk (because the metal is produced domestically in the USA and in many other parts of the world). The rare earths and platinum-group metals, on the other hand, are used in many specific applications where they are difficult to replace, and because they are produced in a small number of not necessarily stable regions, their criticality index is high (Fig. 1.4).

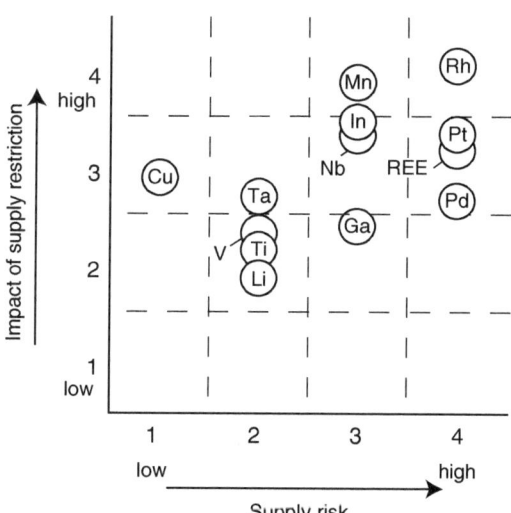

Fig. 1.4 The USGS criticality index

1.4   What Is an Ore Deposit?

An ore deposit is defined *as an accumulation of a useful commodity that is present in high-enough concentration and in sufficient quantity to be extractable at a profit.* In this definition as well we find the terms *useful commodity* and *profit*: the definition is both geological and economic. To understand these ideas, consider the following exercise.

Box 1.4 Selection of a Mining Property
Imagine that you are the director of a mining company and that a prospector comes to you with the following list of properties. You have to decide which is the most attractive target for development in the coming 5–10 years.

1. A deposit of ten million tonnes with 0.2% Cu near Timmins in Canada
2. A deposit of one million tonnes with 1% Cu near Timmins
3. A deposit of ten million tonnes with 2% Cu at Daneborg on the northeast coast of Greenland
4. A deposit of ten million tonnes with 5% Cu in the northeast of Pakistan
5. A deposit of five million tonnes with 1% Cu near Timmins
6. A deposit of 100 million tonnes with 0.7% Cu near Timmins
7. A deposit of 100 million tonnes with 0.7% Cu on the Larzac plateau, France

Response

You see that the list comprises seven hypothetical copper deposits that are distinguished by their size, their grade, and their location. To make a choice, it is easiest to start by eliminating the properties that are the least attractive, either because their size or grade, or because they are located in inhospitable regions. To help with the choice, you will recall that in the previous section we said that the average grade of mined copper is about 0.7% and that a normal deposit contains 100s of millions of tons of ore. With this information we can eliminate deposit number 1, whose grade is too low, and deposit number 2, which is far too small. The deposit at Daneborg, situated on the east coast of Greenland some 500 km north of the Arctic circle, is unattractive because of its small size and its location far from centres of industry in a region with extreme climate*. Given the state of war that exists in the "tribal areas" of northern Pakistan (number 4), no responsible mining company would consider developing a deposit in that region.

This leaves the last three deposits. The Larzac plateau is the home of José Bové, the radical French farmer and professional protestor who rose to fame when he tore the roof off a Macdonald's restaurant. As a passionate anti-capitalist and fierce opponent of the exploration for shale gas, it is most unlikely that he would permit a large copper mine to open on the farm where he produces his roquefort.

The only two that remain, deposits 5 and 6, are in northern Ontario, a region with a long mining history and a political climate favourable to mining. To distinguish between these two we need only consider the amount of copper in each deposit. Deposit 5 contains 50,000 t; deposit 6 contains 700,000 t. The much larger amount of metal in the latter deposit would offset

(continued)

the higher cost of mining its lower grade ore, making deposit 6 the most attractive.

*Were the deposit much larger and on the more hospitable west coast of Greenland, it might be viable. The Black Angel Pb–Zn deposit, located on a precipitous cliff on the margin of a fjord near Maamorilik, was mined successfully from 1973 to 1990 and it closed only because of falling metal prices (Pb and Zn followed a trend similar to that of copper shown as shown in Fig. 1.2). In 2006 the retreat of a coastal glacier revealed another very large and rich Zn-Pb-Ag deposit which is currently being mined – a silver lining on the cloud of global warming?

1.5   Factors that Influence Whether a Deposit Can Be Mined

1.5.1   Tenor and Tonnage

Some idea of the relationship between grade, tonnage and viability of an ore deposit was given in Box 1.4. For a deposit to be mineable it must contain more than a given concentration of the valuable commodity, and more than a given tonnage of this commodity. As shown very diagrammatically in Fig. 1.5a, deposits tend to be distributed along a trend from an extremely small and rich deposit- a single crystal of copper is the extreme example – to another deposit that is very large but with much lower grade – the entire Earth. Most deposits that are both big, close to the surface and high-grade have been mined out and what remains are small rich deposits and much bigger low-grade deposits in more remote regions or at greater depth in the crust.

Figure 1.5b shows in an equally crude way the relationship between grade and price (this relationship is explored in far more detail in a later section). Some metals

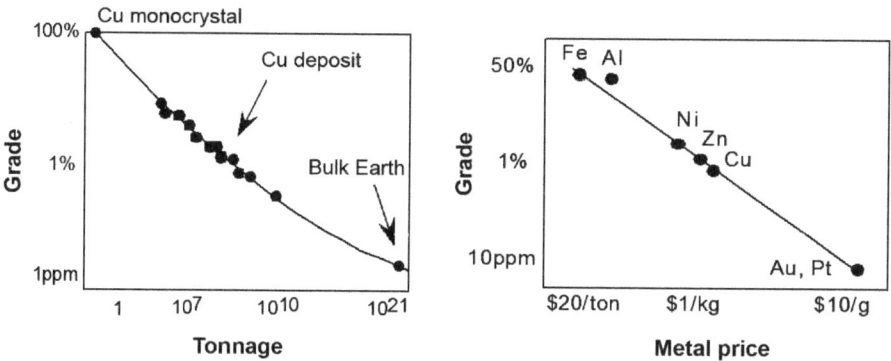

Fig. 1.5 (**a**) Sketch showing variation in the grade and size of ore deposits; (**b**) the relationship between grade and price of selected metals

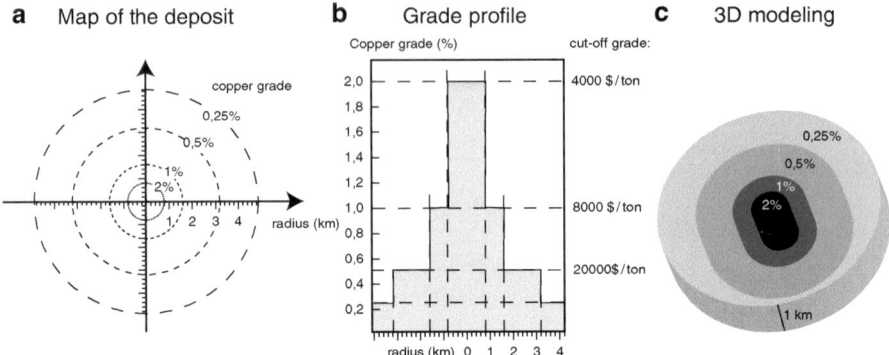

a   Map of the deposit   **b**   Grade profile   **c**   3D modeling

Fig. 1.6 Highly schematized plan of an ore deposit and variations in ore grade, showing how an increase in ore price results in a decrease in the tenor limit which increases the volume of ore that can be mined and the quantity of metal that can be recovered

are abundant in the Earth's crust and they are present in high concentrations in ores. As a consequence their price is relatively low. Other metals are present in far lower concentrations and their price is much higher.

In any deposit the ore type varies, from small areas of rich, high-grade ore to larger areas with lower grades. The values shown in Fig. 1.6a are the average grades that are mined, a mixture of high and low grade ore. What is left in the ground after mining is material, geologically very similar to the material that has been mined, but simply containing a lower concentration of the ore metal, a concentration that is below a certain threshold. This important parameter is called the cut-off grade. To include sub-ore in the material being mined would lead to the operation becoming unprofitable: the cost of mining would exceed the value of the recovered metal.

But what would happen if the metal price improves? It is evident that if the price *increases*, the cut-off grade *decreases* because lower-grade material can then be mined at a profit. As a consequence, the amount of mineable material in the deposit increases.

The example discussed in Exercise 1.5 illustrates clearly how the amount of recoverable metal depends on the price. Taking the argument further, if society requires a commodity, and if no substitute can be found, then the price will increase to the extent that low-grade accumulations of the commodity become ore. There are, of courses, many limits and complications, but this type of argument leads us to suggest that the resources of many or most metals will never be exhausted.

Box 1.5 Estimation of the Amount of Recoverable Ore as a Function of Price and Cut-Off Grade

Figure 1.6 is a sketch of a hypothetical ore deposit: a rich, high-grade core is surrounded by a much larger volume of lower grade material. Suppose that the price of copper increases from \$4,000 to \$8,000 per ton, as it did during the period 2004–2008, and that the increase led to a drop in the cut-off grade.

(continued)

The consequence is that a much larger amount of ore can be mined and the amount of copper that can be recovered increases.

In the example, the radius of the zone that can be mined increases from 1.6 to 2.7 km as the cut-off grade drops from 1% to 0.5%. The tonnage of ore that can be mined depends of the square of this distance (assuming that the maximum depth of mining remains fixed at 1 km) and the volume increases from 2.6 km^3 ($=1.6^2 \times 1$) to 7.3 km^3 ($=2.7^2 \times 1$). Taking into account the lower grade of the newly recoverable ore (0.5 instead of 1%), the tonnage of mined copper almost doubles, from ($2.6 \times 0.01 \times 10^9 =$) 2.6×10^7 to 4.9×10^7 t.

1.5.2   Nature of the Ore

Another factor that strongly influences the viability of a deposit is the nature of the ore. Characteristics to be considered include the type of mineral, the grain size, and the texture of the ore, all of which influence the cost of mining and the extraction of the valuable commodity. The lowest extraction costs are for ores in which the extracted element is only mechanically bound into its gangue (e.g. free-milling gold ores or placer deposits); higher extraction costs are associated with ores in which the element is chemically bound to sulfur or oxide (most base-metal ores) because it takes more energy to break such chemical bonds than to mechanically liberate a particle. The highest extraction costs are for ores in which the element is chemically bound to silicates because these bonds are much stronger than metal-sulfur bonds. Consider, for example, the two major types of nickel ore: magmatic and lateritic. In the first, the ore mineral is sulfide (mainly pentlandite, $(Fe,Ni)_9S_8$) whereas in lateritic ore it is garnierite (a clay mineral) or goethite (Fe hydroxide). Each type of ore has its advantages and disadvantages. The capital investment and the energy required to extract Ni is much higher for the lateritic ores, a major disadvantage in these days of increasing energy costs; on the other hand, the refinement of sulfide ore produces vast amount of sulfur, only some of which can the sold as a by-product.

The grain size and the hardness of the ore influence the cost of grinding it to the fine powder that is fed into the refinery or smelter. Three Zn-Pb ($\pm$Cu) deposits in Australia provide a striking example (Fig. 1.7). All have similar ore grades but the Broken Hill deposit has been metamorphosed to granulite facies and its coarse ore is very easy to process; Mt Isa is less metamorphosed and its finer-grained ore is less attractive; and the virtually unmetamorphosed McArthur River ore is so fine that the ore metals cannot be extracted from waste minerals by simple crushing.

Also to be mentioned in this category are minor elements that increase or decrease the value of an ore. In many cases, the ore contains amounts of valuable metals in concentrations that are below the normal cut-off grade, but if they are extracted as a by-product during the recovery of the major ore metals they

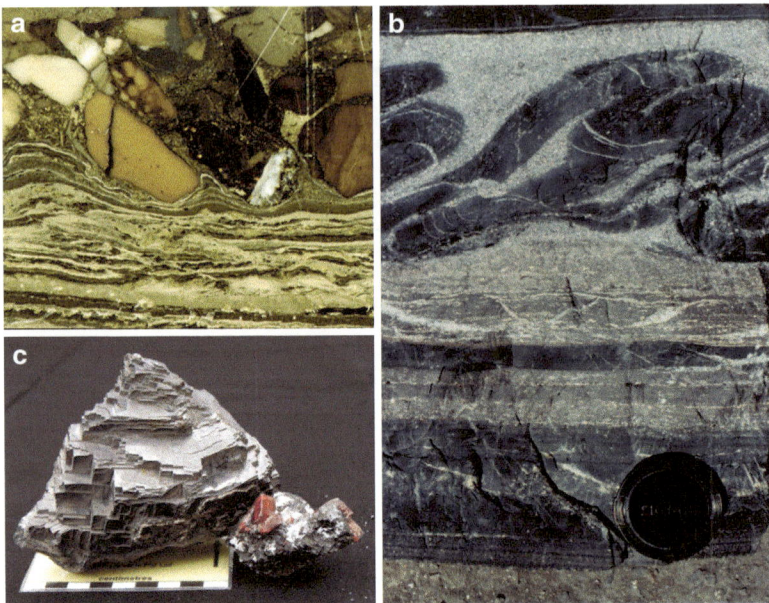

Fig. 1.7 Three types of Pb-Zn sulfide ore distinguished by different grain size. Top left Very fine-grained unmetamorphosed ore from the McArthur River deposit. The pale yellow banded material is fine-grained Zn-Fe sulfides and clay minerals. The detrital grains of quarts and lithic fragments deform these bands; Right Fine-grained and deformed, slightly metamorphosed ore from Mt Isa. Bottom left. Coarse-grained galena and bustamite (Mn-Ca silicate) from Broken Hill where granulite-facies metamorphism has produced in large, easily processed ore (Photo (**a**) from Ross Large, photo (**b**) from Peter Muhling, photo c from Chris Arndt)

contribute significantly to the viability of the operation. Common examples of such "bonus metals" include gold or silver in copper ores, and platinum metals in Ni ores. Another topical example is the rare- earth elements which were initially recovered as a by-product during mining of the Bayan Obo iron deposit in China (see Chap. 6). In contrast, the presence of small amounts of other metals can complicate the extraction process and decrease the value of the ore. Examples of "toxic" or unwanted metals include phosphorous in iron ore and arsenic in base-metal sulfide ores.

1.5.3 Location of the Deposit

In Box 1.4 we also saw the influence of the location of a deposit. Its value, and its very viability, decreases if it is far from centres of industry or population, or in a harsh climate, or in a politically unstable region. All these factors increase the cost of mining or of bringing the metals to market; or they render the operation of a mine too dangerous or risky.

Also important is the geological situation. The largest Ni deposit we know of is in the centre of the Earth. The core contains some 10^{19} tons of Ni metal but it is course totally inaccessible (except for the heroes of American movies). The depth of a deposit has a major influence on the cost of mining. A shallow deposit can be exploited in an open-pit mine, which is far cheaper than the alternative, an underground mine, that must be developed if the deposit is deeper. Friable and soft sedimentary ores are easier to mine and process than ores in hard magmatic rocks. And finally a continuous and compact ore body is far easier to mine than an ore body that is disrupted by faulting or other geological factors. Two platinum deposits in southern Africa provide an interesting example. Those in the Bushveld Complex in South Africa are near-continuous reefs that make the mining operation predictable and efficient, but deposits in another intrusion, the Great Dyke in Zimbabwe, although of similar grade to the Bushveld deposits, are so irregular and disrupted by faulting that mining had proved very difficult. And then the destabilization of the country's economy by the present government has made the operation even more hazardous.

1.5.4   Technical, Economical and Political Factors

As illustrated by the examples discussed above, economic and diplomatic issues may strongly influence the viability of a deposit in some cases increasing its value, in other cases detracting from it. The role of technology, on the other hand, is generally positive. Only through improvements in the techniques used to mine and process ore have we been able to extract metals from deposits with lower and lower grades. One example of this tendency is the decrease in the copper grade discussed at the start of the chapter. Another striking example is the reprocessing of gold ores in Western Australia. The ores of the Coolgardie-Kalgoorlie region were first discovered in 1893 and initially only alluvial gold was exploited. Underground mining soon followed and in the early part of the twentieth century, vast waste dumps from underground mining littered the surroundings of the growing boomtowns. In the following century these dumps have been reprocessed three or four separate times and each time gold that had previously been discarded was recovered. The process was driven by sudden increases in the price of gold, notably with the abandonment of the gold standard in 1971 and the more recent hike in the gold price associated with the metals boom at the start of this century. But coupled with these economic pressures were technological advances that allowed the recovery of gold that was unattainable using earlier techniques. The most recent involves in-situ leaching in which fluids, commonly containing gold-eating bacteria, are allowed to percolate through the waste dumps. Other advances include the development of more efficient mining methods, as best expressed in the vast open cast mines that exploit large, low-grade, near-surface deposits of copper, gold, iron and other metals.

Finally, global and local economic and political situation can strongly influence the viability of a deposit, as illustrated in the examples described earlier in the chapter.

References and Further Reading

British Geological Survey (2010), World mineral production (2005–2009) http://www.bgs.ac.uk/mineralsuk/statistics/worldStatistics.html

Malthus TR (1830) An essay on the principle of population. Penguin Classics London. ISBN 0-14-043206-X

Meadows DH, Meadows DL, Randers J, Behrens WW (1972) The limits to growth. New York. Universe Books, Universe Books, New York, p 207 pp

Mudd GM (2010) The environmental sustainability of mining in Australia: key mega-trends and looming constraints. Resour Policy 35(2):98–115

National Research Council (2008) Minerals, critical minerals, and the U.S. economy. The National Academies Press, Washington. ISBN 0309112826, 264 p

United States Geological Survey (2010), Mineral Resources Program. http://minerals.usgs.gov/products/index.html

Chapter 2
Classification, Distribution and Uses of Ores and Ore Deposits

2.1 Classifications of Ores

The geological literature contains many schemes for classifying ore minerals. Some have an economical basis linked to the end use of the metal or mineral that is derived from the ore; others depend partially or entirely on geological factors.

2.1.1 Classifications Based on the Use of the Metal or Ore Mineral

In older books it is common to find minerals classified, as in Table 2.1, according to the use that is made of the metal or mineral extracted from the ore. For example, Table 2.2, a classification of ore minerals, contains some of the minerals that are mined for copper. We see that this metal is extracted from various types of sulfides (e.g. covellite) and sulfosalts (tetrahedrite), as well as from carbonates (malachite), oxides (cuprite) and in rare cases as a native metal. Copper is one of the "base metals", a term that refers to a group of common metals, dominated by the transition elements, which are widely used in industry. Gold and platinum are classed as "precious metals". Other classes of ores comprise minerals that are used in their natural state without refinement or extraction of a specific element. Barite, a sulfate of the heavy element barium, is employed to increase the density of the fluids ("muds") used when drilling for oil. Uranium and coal are sources of energy. Various types of hard minerals are used as abrasives; garnet and industrial diamond are two examples, as is feldspar (next time you buy a tube of cheap toothpaste, see if it contains "sodium-aluminium silicate"). This type of table provides a useful link between the various types of ores and the use that society makes of them.

N. Arndt and C. Ganino, *Metals and Society: an Introduction to Economic Geology*,
DOI 10.1007/978-3-642-22996-1_2, © Springer-Verlag Berlin Heidelberg 2012

Table 2.1 Metals, useful minerals and their ores

Class	Element	Mineral	Composition
Ferrous metals	Iron (Fe)	Hematite	Fe_2O_3
		Limonite, goethite	FeO.OH
		Magnetite	Fe_3O_4
	Manganese (Mn)	Pyrolusite	MnO_2
	Chromium (Cr)	Chromite	$FeCr_2O_4$
	Nickel (Ni)	Pentlandite	$(Fe,Ni)_9S_8$
		Garnierite	$(Ni,Mg)_3Si_2O_5(OH)_4$
	Molybdinum(Mo)	Molybdenite	MoS_2
	Vanadium (V)	Magnetite	$(Fe,V)_3O_4$
Aluminium	Aluminium (Al)	Gibbsite	$Al(OH)_3$
Base metals	Copper (Cu)	Chalcopyrite	$CuFeS_2$
		Chalcocite	Cu_2S
		Cuprite	Cu_2O
		Tetrahedrite	$(Cu, Ag)_{12}Sb_4 S_{13}$
		Malachite	$CuCO_3.Cu(OH)_2$
		Azurite	
		Native copper	Cu
	Zinc (Zn)	Sphalerite	(Zn, Fe)S
	Lead (Pb)	Galene	PbS
	Tin (Sn)	Cassiterite	SnO_2
Precious metals	Gold (Au)	Native gold	Au
	Platinum (Pt)	Alloys of platinum group elements (PGE)	Pt, Pd, Os, Ir
	Silver (Ag)	Native silver	Ag
Argentite (Ag_2S)	Ag		
Energy sources	Uranium (U)	Pitchblende	UO_2
	Coal (C)	Coal	C
High-technology metals	Titanium (Ti)	Imenite	$FeTiO_3$
	Zirconium (Zr)	Zircon	$ZiSiO_4$
	Niobium (Nb), thorium (Th), rare earth elements	Monazite, apatite and rare minerals (bastnäsite, pollusite, etc)	Nb, Th, La, Ce, Nd
Lithium (Li)			
Beryllium (Be)			
Spodumene			
Li-rich brine			
Lepidolite beryl ($LiAlSi_2O_6$)			
$Be_3Al_2(SiO_3)_6$			
Other elements	Barium (Ba)	Barite	$BaSO_4$
	Fluorine (F)	Fluorite	CaF_2
	Potassium (K)	Sylvite	KCl
Minerals		Diamond	C

(continued)

Table 2.1 (continued)

Class	Element	Mineral	Composition
		Corundum	Al_2O_3
		Grenat	Silicate of Al, Mg, Fe
		Talc	Phyllosilicate
		Mica	Phyllosilicate
		Diatomite and	Clay
		Alusite, kyanite	Al_2SiO_5
		Albite	$NaAlSi_3O_8$
		Halite	NaCl
		Calcite	$CaCO_3$

2.1.2 Classifications Based on the Type of Mineral

The type of mineral provides the basis of classification given in Table 2.2. Here we see that a wide range of important metals are mined in the form of sulfide (e.g. Cu as chalcopyrite, Pd as galena, Ni as pentlandite). Another important class are oxides, which are mined for tin, as in the mineral cassiterite (SnO_2), or Fe in magnetite (Fe_3O_4) and uranium in pitchblende (UO_2). Other types of metals are found as carbonates or sulfates, usually in alteration zones overlying primary deposits.

Very few metals are mined in their native form, the only common examples being gold and the platinum-group elements. Carbon is also mined as a native element as diamond or graphite, and in an impure form as coal. Although copper does occur as a native metal, its occurrence in this form is usually more an impediment than an advantage; although native copper does contain 100% Cu and its presence boosts the copper grade, the mineral in malleable and tends to gum up the crushing machines which are designed for brittle sulfides and silicates.

Silicates, by far the most important rock-forming mineral, are uncommon in the list of ore minerals. Exceptions are garnierite, a clay-like mineral that is the major ore mineral in Ni laterites; zircon ($ZrSiO_4$), a heavy detrital mineral mined for the high-technology metal zirconium; and garnet, which is used as an abrasive. Quartz is becoming increasing important as a source of the silica that is used in semiconductors and in solar panels.

> **Box 2.1 Copper, a Highly Versatile Metal**
> Copper, along with gold, was one the first metals to be used by mankind and it is very widely used today. It is mined in almost all parts of the world, and it is used very widely in industry. The major copper producing countries are Chile, USA, Peru and China. Almost every country is a consumer of copper, the level depending on the size of the population and the extent of industrialization.
>
> *(continued)*

Table 2.2 Classification of ore minerals

Sulfides and sulfosalts

Covellite – CuS

Chalcocite - Cu_2S

Chalcopyrite - $CuFeS_2$

Bornite – $Cu_8 FeS_4$

Tetrahedrite – $(Cu, Ag)_{12}Sb_4 S_{13}$

Galena - PbS

Sphalerite – $(Zn,Fe)S$

Cinnabar - HgS

Cobaltite - $(Co, Fe)AsS$

Molybdenite: -MoS_2

Pentlandite: $(Fe, Ni)_9S_8$

Millerite – NiS

Realgar – AsS

Stibnite – Sb_2S_3

Sperrylite – $PtAs_2$

Laurite – RuS_2

Oxysalts

Calcite – $CaCO_3$

Rhodochrosite – $MnCO_3$

Smithsonite – $ZnCO_3$

Malachite – $Cu_2(OH)_2CO_3$

Barite - $BaSO_4$

Gypsum – $CaSO_4 .2H_2O$

Scheelite - $CaWO_4$

Wolframite - $(Fe, Mn)WO_4$

Apatite – $Ca_8 (PO_4)_3 (F,Cl, OH)$

Halides

Halite – NaCl

Sylvite – KCl

Fluorite – CaF_2

Oxides and hydroxides

Bauxite

  Gibbsite - $Al(OH)_3$

  Boehmite - $(\gamma\text{-}AlO(OH))$

  Diaspore - $(\alpha\text{-}AlO(OH))$

Cassiterite - SnO_2

Cuprite – Cu_2O

Chromite - $(Fe, Mg)Cr_2O_4$

Columbite -tantalite or coltan

  - $(Fe, Mn)(Nb, Ta)_2O_6$

Hematite - Fe_2O_3

Ilmenite - $FeTiO_3$

Magnetite - Fe_3O_4

Pyrolusite -MnO_2

Rutile – TiO_2

Uraninite (pitchblende) - UO_2

Metals and native elements

Gold – Au

Silver – Ag

Platinum-group metals – Pt, Pd, Ru

Copper - Cu

Carbon – C (diamond, graphite)

Silicates

Beryl - $Be_3 Al_2 (SiO_3)_6$

Garnet – $Fe_3 Al_2 (SiO_4)_3$

Garnierite – mixture of the Ni-Mg-hydrosilicates

Kaolinite – $Al_4 Si_4 O_8 (OH)_8$

Sillimanite – $Al_2 SiO_8$

Spodumene – $LiAlSi_2 O_6$

Talc – $Mg_3 Si_4 O_8 (OH)_2$

Zircon – $ZrSiO_4$

Common uses of copper are given in the following table. Its high electrical and thermal conductivity, its resistance to corrosion and its attractive colour lead to a vast range of applications. It is used as wire to conduct electricity in electrical appliances of all types and in alloys with zinc (brass) or other metals in utensils and coins. The development of new types of alloys has led to new uses in superconductors and batteries; and copper compounds are used in a wide variety of products such as pesticides (copper sulfate pentahydrate controls fungus on grapes and algae in swimming pools) and in antibiotics (Table 2.3).

Table 2.3 Uses of copper
(Source: Standard CIB Global Research www.standardbank.co.za)

Electricity, electronics	42%
Construction	28%
Transport	12%
Industrial machinery	9%
Other (coins, medicines, fungicides)	9%

In developed countries, the per capita consumption of copper has remained near constant of the past decades. The new uses of the metal generally require only relatively small quantities and these additions have been countered to some extent by the abandonment of some large-scale industrial applications and by increased recycling. However, increasing demand from developing countries will require that global production be increased significantly; this production can only be met by the discovery of new deposits and the efficient exploitation of these deposits. In addition, with the increasing use of electronics in cars and household or industrial devices, per-capita use of copper in wiring and circuits, in both developed and developing countries, is expected to increase.

2.2 Classifications of Ore Deposits

There are some parallels between the schemes used to classify ores and those used to classify ore deposits. Again in older texts, deposits are classified according to the type of product they produce; copper deposits, gold deposits, energy sources (uranium and coal), and so on. This type of classification finds some application in a purely economic context but is not employed here.

Through the twentieth century many classifications were based on the types of rocks that host the ore deposit or on the geometry of the deposit and its relation to the host rocks. An example is given in Table 2.4. Deposits in granites were distinguished from those in sedimentary rocks; vein-like deposits were distinguished from layers conformable with the stratification of the host rock; massive ores were distinguished from disseminated ores. A popular classification developed by Lindgren, an influential American economic geologist, distinguished deposits that formed at different levels in the crust (Table 2.4). The terms "epizone", "mesozone" and "catazone", for deposits at shallow, intermediate, and deep levels in the crust, are still employed today. A further distinction was made between "syngenetic" deposits, which formed together with and as part of the host rock, and "epigenetic" deposits, which formed through introduction of ore minerals into already consolidated rocks.

The development in the latter part of the twentieth century of the theory of plate tectonics spawned a swarm of classifications based on tectonic settings. As shown in Table 2.5, deposits in ocean basins were distinguished from those in convergent margins or intracratonic settings, and so on. This type of classification is still used,

Table 2.4 Lindgren's classification of ore deposits (Modified from Lindgren 1933 and Evans 1993)

	Depth	Temperature (°C)	Occurrence	Metals
Telethermal	Near surface	± 100	In sedimentary rocks or lava flows; open fractures, cavities, joints. No replacement phenomena	Pb, Zn, Cd, Ge
Epithermal	Near surface to 1.5 km	50–200	In sedimentary or igneous rocks; often in fault systems; simple veins or pipes and stockworks; little replacement phenomena	Pb, Zn, Au, Ag, Hg, Sb, Cu, Se, Bi, U
Mesothermal	1.2–4.5 km	200–300	Generally in or near intrusive igneous rocks; associated with regional faults; extensive replacement deposits or fracture fillings; tabular bodies, stockworks, pipes	Au, Ag, Cu, As, Pb, Zn, Ni, Co, W, Mo, U etc
Hypothermal	3–15 km	300–600	In or near deep-seated felsic plutonic rocks in deeply eroded areas. Fracture-filling and replacement bodies; tabular or irregular shapes	Au, Sn, Mo, W, Cu, Pb, Zn, As

Table 2.5 Tectonic classification of ore deposits

I. Deposits at oceanic ridges (divergent plate margins)
Volcanogenic massive sulfide deposits (Cu, Zn)
Exhalative deposits (Zn, Cu, Pb, Au and Ag). e.g. Red Sea
Mn nodules (Mn, Ni, Cu, Co)
Cr, PGE, asbestos in ultramafic rocks

II. Deposits at convergent plate margins
Porphyry Cu-Mo deposits
Other base metal deposits (Cu, Pb, Zn, Mo).
Precious metals (Pt, Au, Ag).
Pb-Zn-Ag veins and contact metasomatic deposits
Other metals (Sn, W, Sb, Hg).

III. Deposits in cratonic rift systems
Deposits of Sn, fluorite, barite in granites
Evaporites in rift basins
Carbonatites containing Nb, P, REE, U, Th and other rare elements

IV. Deposits in intracontinental settings
Ni and PGE in layered intrusions
Ti in anorthosites
Iron-oxide Cu-Au deposits
Pb-Zn-Ag deposits in limestones and clastic sediments
Sedimentary Cu deposits
Ni, Al laterites
Diamonds in kimberlites

particularly when discussing the broad-scale distribution of ore deposits, as we do in the following section. However, newer schemes in which the basic criterion is the *ore-forming process* have largely replaced this type of classification (Table 2.6). Although it might be argued that a rigorous classification should be based on objective parameters that can be measured and quantified, and not on properties that must be inferred, this is the classification we will use in this book.

The scheme we have chosen has some disadvantages, and, as will be seen in following chapters, it is sometimes not at all clear whether a certain deposit should be placed in one box and not another, but it also has the great advantage that it emphasizes that an ore deposit results from a normal geological process like those that form common igneous or sedimentary rocks. It provides an incentive to move the discipline from "gitologie" – a French term that can the translated as "depositology", a descriptive catalogue of ore deposits – to a modern interpretative science. Finally, the approach provides a means of applying knowledge of geological processes including concepts such as the partitioning of major and trace elements between melt and crystal, the sorting of light from heavy minerals during fluvial transport, or the stability of mineral phases in aqueous solutions, to develop an understanding of how an ore deposit is created.

2.2.1 A Classification Based on the Ore-Forming Process

The list of headings in Table 2.6 overlaps the list of geological processes that would be found in any introductory geology text. We see for example that magmatic processes form some deposits, and sedimentation or surface weathering form

Table 2.6 Genetic classification of ore deposits

1. Magmatic: ores that form by the accumulation of minerals that crystallize directly from magma.
  (a) In mafic and ultramafic rocks
    • Chromite and platinum-group elements (PGE) in large layered intrusions (Bushveld in South Africa, Great Dyke in Zimbabwe)
    • Chromite in ophiolites (Turkey)
    • Cu-Ni-Fe sulfide in the layered intrusions (Sudbury, Noril'sk)
    • Sulfide Ni-Cu-Fe in komatiitic lavas (Kambalda)
    • Diamonds in kimberlites
  (b) Associated with felsic intrusions
    • Cu ore in carbonatites (Phalabora)
    • REE, P, Nb, Li, Be etc in pegmatites
2. Deposits associated with hydrothermal fluids: metals are mobilized within and precipitated from hot aqueous fluids or various origins
  (a) Cu-Mo-W deposits in granitic intrusions
    • Deposits of the type "porphyry" (porphyry copper) (USA, Chile, Philippines)
  (b) Epigenetic deposits – minerals in veins or replacing host rocks
    • Cu, Zn, Pb, Mo, Ag, Au ores related to granitic rocks (Butte, Potosi)
    • "orogenic" gold deposits (Abitibi, Canada, Yilgarn, Australia)
    • Ag-Ni-Co-As-S deposits (Potosi, Bolivia; Cobalt, Canada)
  (c) Volcanogenic massive sulfide (VMS) deposits
    • Precambrian deposits (Noranda, Canada)
    • Modern (Kuroko, Japan)
  (d) Deposits unrelated to magmatic activity
    • Pb-Zn ore in limestone (Mississippi Valley Type, USA)
    • Uranium deposits (Athabasca, Canada; Colorado, USA)
3. Sedimentary deposits: concentrations of detrital minerals or precipitates
  (a) SEDEX – Pb-Zn sulfides in shales (Mt Isa, Australia; Sullivan, Canada)
  (b) Cu ores in sandstones (Copperbelt of Central Africa; Kupferschiefer, Poland)
  (c) BIF (banded iron formations) (Australia, Brazil, Canada)
  (d) Evaporites, phosphatites, Li-rich brines, limestone
  (e) Placer deposits
    • Placer gold in rivers (California, Australia, Brazil)
    • Ti, Zr in beach sands (Australia)
    • Diamonds in sand and gravel (South Africa)
4. Deposits related to weathering
  (a) Al laterites – bauxite (Jamaica, France, Australia)
  (b) Ni laterite (New Caledonia)
  (c) "Supergene" ore enrichment
5. Metamorphic deposits
  (a) Deposits in skarn (China, Scandinavia, USA)

others. What distinguishes the two lists is the minor importance of metamorphism in the list of important ore-forming process and, in its place, a major class comprising deposits that are linked to hydrothermal fluids. When the operation of the Earth as a whole is considered, the circulation of hot aqueous fluids through the crust is normally mentioned only as an agent that alters the composition or texture of primary magmatic or sedimentary rocks; the same process, however, lies at the origin of a vast range of important ore deposits and has created well over half of all ore bodies that are known to exist.

Consider now the first category in Table 2.6, magmatic deposits. We note that many types of large and important deposits are found in mafic-ultramafic rocks and only a few less important types in evolved, felsic or silicic, rocks. Many ore deposits are indeed hosted by granites, but according to modern ideas of ore genesis, such deposits generally result from the precipitation of ore minerals from aqueous fluids and not from the granitic magma itself. The type of ore mineral in magmatic deposits is directly linked to the composition of the host rock. In mafic-ultramafic hosts we find deposits of Ni, Cr and platinum-group elements, all of which partition strongly into minerals that crystallize early in normal magmatic differentiation. Ores in felsic rocks, by contrast, are confined to elements that concentrate in evolved magmatic liquids. Some of these are present in late-crystallizing phases such as ilmenite, which contains Ti, and cassiterite, the ore of Sn; others enter the water-rich fluid that separates from the silicate liquid, to be redeposited in pegmatites or in hydrothermal ore bodies. Pegmatites are important sources of rare but increasingly important metals such as Li and Be.

Some metals are restricted to a single type of ore-forming process, the best example being Cr, which, with virtually no exceptions, is mined as chromite, a magmatic oxide that accumulates during the crystallization of mafic or ultramafic magmas. Some very minor placer (sedimentary) chromite deposits have been mined, but hydrothermal deposits are unknown. More than 98% of Al is mined as bauxite, a lateritic soil that forms in hot and humid climates; but the same metal is also extracted in a Russian mine in nepheline syenite, a magmatic rock. Most metals, however, occur in deposits of diverse origins. Alloys of the platinum-group elements and cassiterite, an oxide of tin, are cited as type examples of magmatic ores, but when the magmatic host rocks are exposed to erosion at the surface, the same minerals may become re-concentrated by fluvial processes to form sedimentary placer deposits. Metals such as copper and gold are represented in almost every class of deposit in the list. The distribution of ore metals and the types of process that create their deposits are discussed in more detail in following chapters.

2.3   Global Distribution of Ore Deposits

Ore deposits are not distributed uniformly across the globe. Vast tracts of land are devoid of viable deposits while others constitute what is known as a 'metal province', a region containing an unusually high proportion of deposits of one or several different types. Notable examples include the string of enormous copper deposits along the American Cordillera from Alaska to Chile, the clusters of lead-zinc deposits in limestones of the central USA, and the tin-bearing granites of SE Asia. For both geological and economic reasons it is important to have some knowledge of this distribution. From a geological point of view, the distribution provides important clues to the ore-forming process; from an economic point of view, the irregular distribution strongly influences metal prices and global trade, and lies at the heart of many of the alliances and conflicts that govern relationships between countries around the world.

In plate tectonic classifications of ore deposits, the emphasis is quite naturally placed on the tectonic setting in which the deposit occurs, but many deposits develop in sedimentary settings or as a result of superficial weathering; in such cases geomorphology, surface relief and modern or past climate exert an additional important influence of the localization of the deposits. All these factors are discussed briefly in the following section and are then elaborated upon in subsequent chapters.

2.3.1   Geological Factors

The global distribution of ore deposits is illustrated in the series of maps of major deposits that are compiled in Fig. 2.1. We have selected only a few major commodities that serve to illustrate the basic principles that govern the distribution of ores; more detailed and exhaustive information is found in standard texts and on the internet, as listed at the end of the chapter.

We will start with copper, an industrial metal that is used in every country and is mined in all parts of the world (Fig. 2.1a). A large proportion of the resources of this metal is tied up in a single type of deposit, the so-called "porphyry copper" or simply "porphyry" deposits (Chap. 4). These deposits are directly associated with subduction zones and thus are found in island arcs and convergent margins. This is the reason for the string of deposits that extends not only along the entire west coast of North and South America, but also throughout the islands of the southwest Pacific. Large deposits of the same type are also found in accreted island arcs that have been incorporated into continental collision zones, as in the Alpine-Carpathian-Himalayan belt. Another major class of copper deposits form within mature sedimentary rocks in intracratonic basins, as in the deposits of the central African "copper belt". Copper also occurs in deposits associated with volcanic rocks, the VMS deposits of Fig. 2.1b, and in deposits associated with shales, the SEDEX deposits in Fig. 2.1b. The metal is also mined in magmatic intrusions, most of which form in intracratonic settings (Fig. 2.1c). Two important examples formed in very different ways. The Norilsk deposits in northern Siberia are associated with a large igneous province and those of the Sudbury area in Canada formed during crystallization of a melt sheet created by the impact of a large meteorite. Another unusual example of a copper ore body is the Palabora

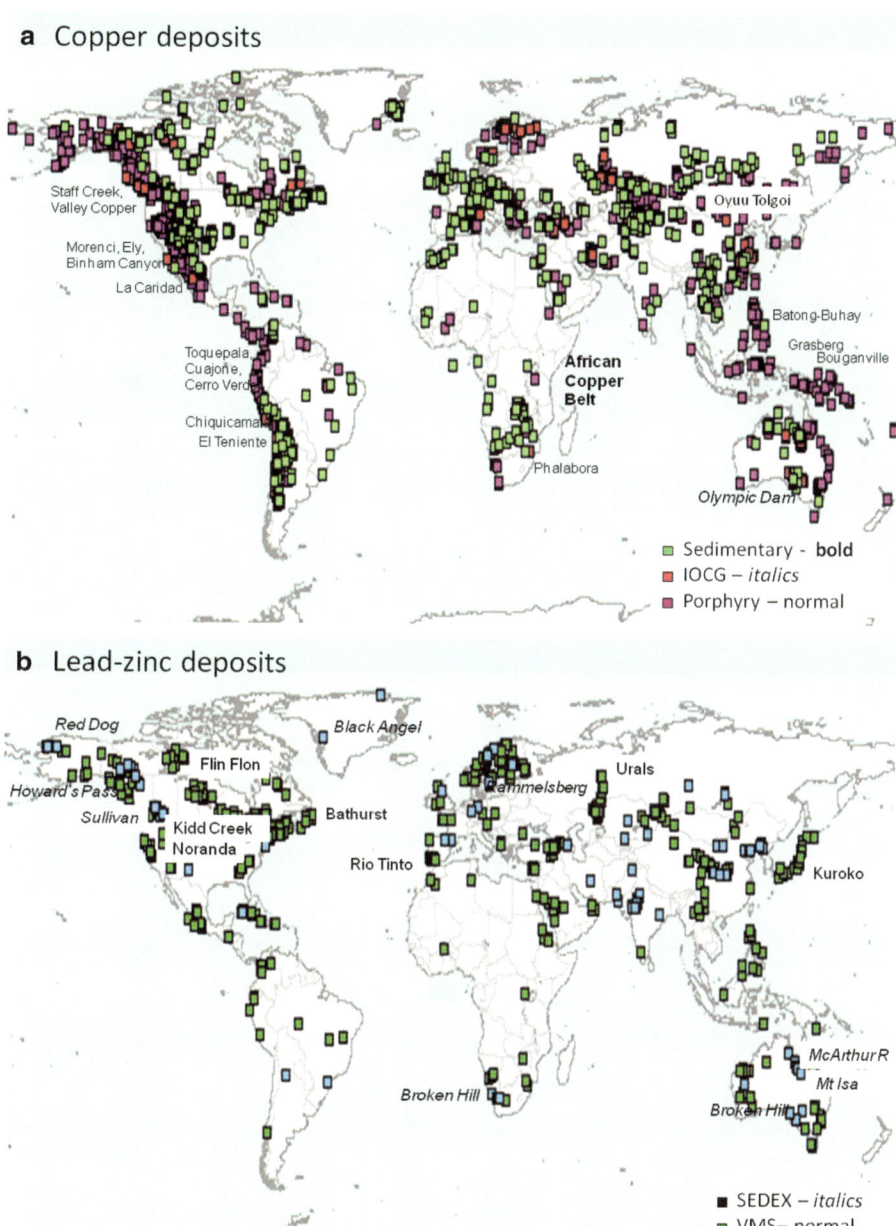

Fig. 2.1 (continued)

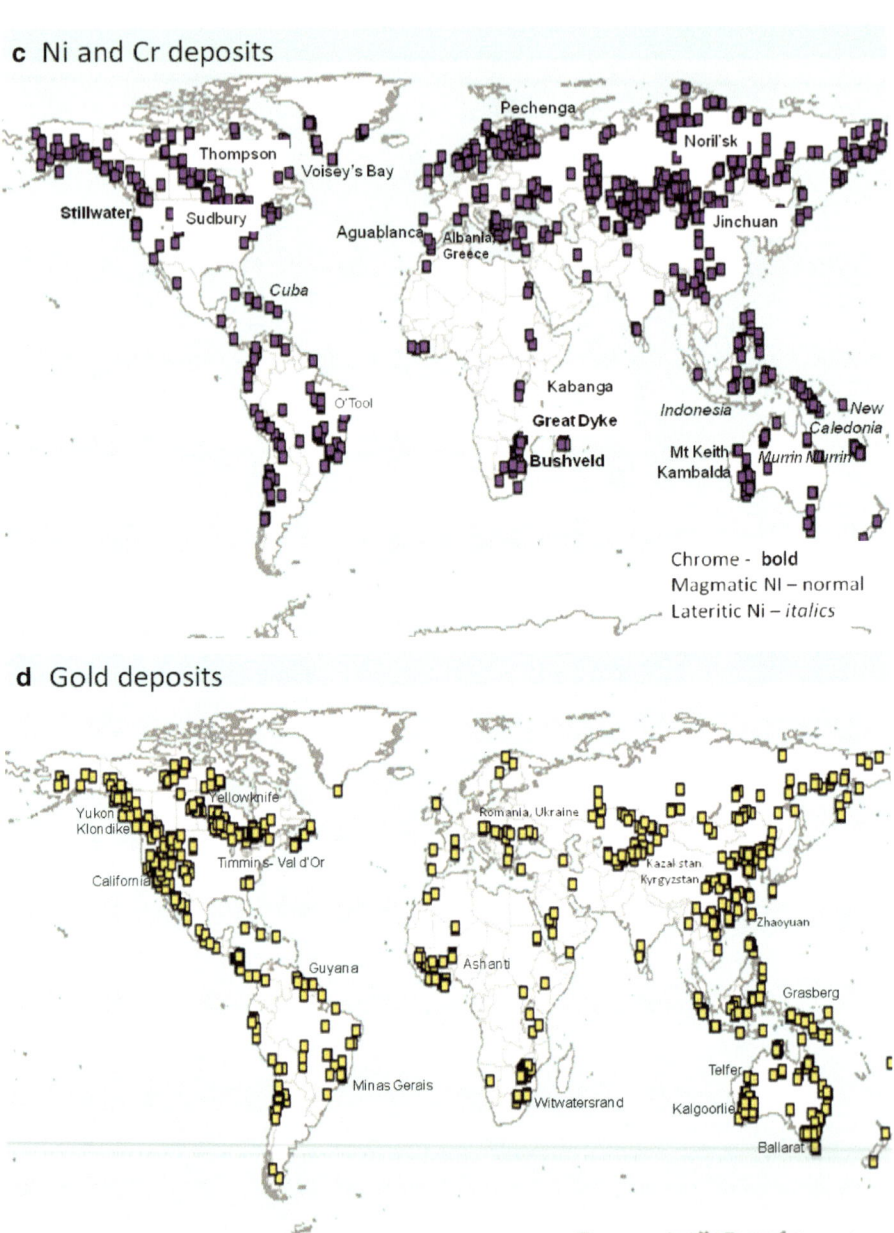

c Ni and Cr deposits

Pechenga

Noril'sk

Thompson

Voisey's Bay

Stillwater   Sudbury

Jinchuan

Aguablanca

Albania Greece

Cuba

O'Tool

Kabanga

Indonesia

New Caledonia

Great Dyke

Mt Keith   Murrin Murrin

Bushveld   Kambalda

Chrome - **bold**
Magmatic NI – normal
Lateritic Ni – *italics*

d Gold deposits

Yukon Klondike   Yellowknife

Romania, Ukraine

Kazakstan Kyrgyzstan

California   Timmins- Val d'Or

Zhaoyuan

Guyana   Ashanti

Grasberg

Telfer

Minas Gerais   Witwatersrand   Kalgoorlie

Ballarat

Fig. 2.1 (continued)

e Uranium deposits

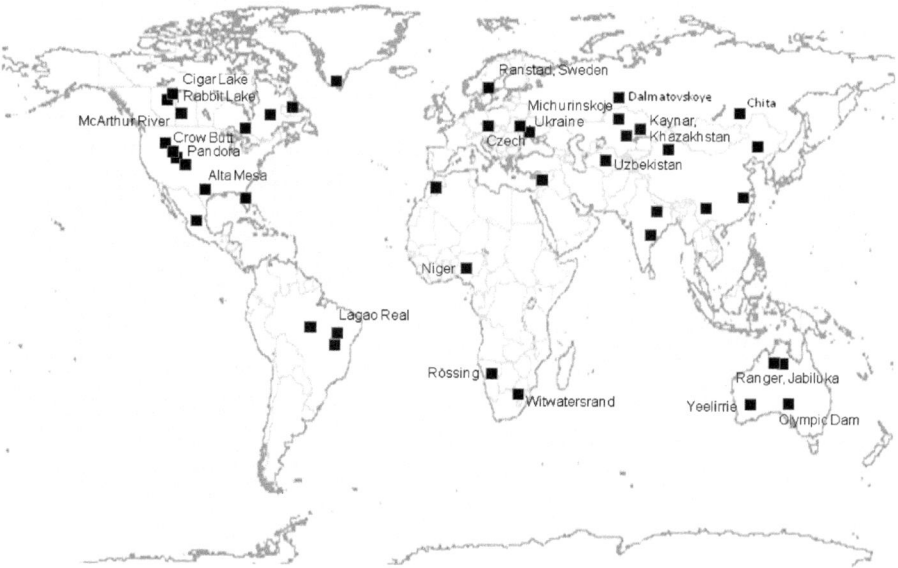

Fig. 2.1 Global distribution of ore deposits. (**a**) Copper deposits. (**b**) Lead-zinc deposits. (**c**) Ni and Cr deposits. (**d**) Gold deposits. (**e**) Uranium deposits from http://gdr.nrcan.gc.ca/minres/index_e.php

(also spelled Phalabora) intrusion, a carbonatite emplaced in the Archean Kaapvaal craton of South Africa.

In many magmatic deposits, copper occurs together with nickel (Fig. 2.1c). This is the case for most major deposits of this type, not only Sudbury but also Norilsk in Russia and Jinchuan in China. Another class of nickel deposit is hosted by komatiite, a type of ultramafic lava that erupted only in the Archean and early Proterozoic. Komatite-hosted Ni-Cu deposits are therefore restricted to the oldest parts of the earth's crust, in the greenstone belts of Australia, Canada and Zimbabwe. But not all nickel deposits are magmatic; another major type is nickel laterite and for these the distribution is quite different. Whereas crustal structure and tectonic setting influence the location of the magmatic variety, laterite is a type of soil that develops at the surface of the crust in hot, humid climates. All deposits of this type are located in regions that are relatively close to the equator, or were close to the equator when the deposits formed. The major lateritic Ni deposits are in New Caledonia, Indonesia, Cuba, Brazil and Australia.

Gold deposits form in a wide range of tectonic settings. The majority of those shown in Fig. 2.1d are classed as orogenic gold deposits, which means that they occur in active margins of the continents, as, for example, the Cordillera of North America or the Alpine-Himalaya belt. Other examples occur in modern island arc settings and in the western Pacific or in their Precambrian equivalent, the Archean greenstone belts of Canada and Australia. Quite a different form of gold deposit is

Box 2.2 The Mining and Refining of Nickel Ores
Mining

Nickel is found in two different types of ore-, magmatic sulfide and lateritic. The former are usually mined by underground techniques or in large and deep open pits for some new deposits; the latter are mined in shallow pits using heavy earth-moving equipment such as shovels, draglines, and front-end loaders The ore is usually ground to coarse aggregate and then it is transported to the refinery.

Refining and Smelting

The ore is first ground in large mills to powder that is fine enough that the particle size is less than that of individual grains of the ore minerals. Sulfide grains are separated from the gangue by the floatation process. The ground ore is mixed in large vats with water and chemicals such as fatty acids and oils that increase the hydrophobicity of the sulfide particles. Mechanical and pneumatic devices stir the mixture and produce air bubbles which are injected at the base of the vats. Sulfide particles adhere to the bubbles and float to the surface where they are recovered by scraping away the froth. The magnetic properties of Fe-Ni sulfides are used to aid their extraction.

The nickel concentrates are then dried, mixed with flux, and heated to about 1,350°C in an oxidizing environment in smelters. The reaction of oxygen with iron and sulfur in sulfide ore supplies some of the heat required for smelting. The product is an artificial nickel-iron sulfide known as matte, which contains 25–45% nickel. The iron is then converted to an oxide, which combines with silica flux to form a slag. When the slag is drawn off, the matte contains 70–75% nickel.

The nickel matte is either leached at high pressure with ammonia and the metal is recovered from solution, or the matte is roasted to produce high-grade nickel oxide. The final stage is electrorefining: the nickel oxide is dissolved in sulfate or chloride solutions in electrolytic cells and pure nickel metal is deposited on the cathode. Sulfur is released in large quantities at several stages of the process. Some is recovered to be used in industry or as a fertilizer, but a large fraction is lost in smelter fumes and constitutes a serious pollutant.

Laterite nickel ores do not contain sulfur and cause less of a pollution problem, but their refinement requires much higher energy input. The ore minerals are oxide or silicates that are not amenable to floatation and other conventional processes. Large tonnages or untreated ore must therefore be smelted. In addition, the reactions of oxide ores from laterite deposits are not exothermic and this increases the amount of energy required by the smelter. Hydrous ore and gangue minerals have high water contents which is removed by drying and roasting in large high-temperature kilns. The nickel oxide is then smelted in furnaces that run at $1,360°C–1,610°C$, the high temperatures being required to accommodate the high magnesium content. Most laterite smelters produce a ferronickel alloy that is sold directly steel manufacturers.

the placers, including the huge early Proterozoic example of the Witwatersrand in South Africa and the more modern placers of California, Victoria, the Klondike and the Yukon.

The final mineral in our short selection of commodities is uranium (Fig. 2.1e). An important class of deposits groups those localized at unconformities at the base of Proterozoic sedimentary basins on Archean cratons in northern Canada and northern Australia. Hydrothermal deposits in the USA and through central Asia occur in younger sedimentary basins. Two notable examples where uranium is produced in multi-element deposits are Olympic Dam in Australia (Fig. 2.1a) and the Witwatersrand conglomerates of South Africa. The Rössing deposit in Namibia is magmatic, Randstad in Sweden occurs in black shales, and the Yeelirrie deposit of Australia is hosted by surficial sediments (calcretes).

Most major iron deposits (not shown in Fig. 2.1) formed in a very specific geological setting during a unique period of Earth history. About 90% of iron ore is mined from "banded iron formations" or BIF, a type of chemical sediment that precipitated from seawater on shallow continental platforms during the early Proterozoic. As explained in Chap. 5, this period in Earth history saw a marked increase in the oxygen content of the atmosphere and oceans, a process that triggered the precipitation of iron oxide that had been dissolved in seawater. Most of the world's great iron deposits are therefore found in sedimentary sequences overlying Archean cratons; in Brazil, Australia, Canada and Russia.

Titanium is mined in two very different types of deposit. The most common ore is ilmenite, a mineral that occurs as an accessory phase in a wide variety of igneous and metamorphic rocks but also in much higher abundances in a special type of rock called anorthosite. This rock consists essentially of calcic plagioclase with a few percent of ferromagnesian minerals and a variable amount of Fe-Ti oxide. A specific type called "massif anorthosite" was emplaced in continental crust during the middle Proterozoic and this type commonly contains mineable concentrations of ilmenite. Large deposits of this type are found in a belt that extends from Quebec in Canada through to Norway.

When igneous or metamorphic rocks are subjected to chemical weathering and erosion, ilmenite, which is stable under these conditions, is released, transported in rivers and re-deposited at the coastline. When the continental crust is stable and subjected to protracted periods of weathering, and when the coastline is a stable passive margin, then large accumulations of dense stable minerals may build up in beach sands. Major deposits of ilmenite, together with associated heavy minerals such as rutile (another source of Ti), zircon (a source of Zr) and monazite (a source of Th and the rare earth elements) occur in sands along the coasts of Australia, India and South Africa.

Diamond also in mined in two types of deposit. The major and primary source is kimberlite, a rare type of ultramafic rock that is emplaced as pipe-like intrusions at or near the margins of Archean cratons. And when kimberlite is eroded, the diamonds are released and accumulate in alluvial deposits in rivers or in sands and gravels at the coast. In historical times most diamonds were found in alluvial deposits in India (from a source whose location that remains unknown). Then the major deposits in South Africa, both kimberlitic and alluvial, were discovered and these provided the bulk of mined diamond for most of the twentieth century. In the last decade, new deposits have been located in almost every country with a stable Archean craton; in Russia, Australia, Canada, Brazil, Greenland and Finland.

2.4　Global Production and Consumption of Mineral Resources

The lists in Tables 2.7 and 2.8 indicate where metals and ore minerals are mined and consumed. The first tables ranks countries in terms of their production of mineral resources; the second the amount of the commodity they consume. Note that petroleum is not included in the selection of commodities. Although many of the largest countries figure near the top of each list, as is to be expected, there are a number of anomalies that provide useful information about how the global minerals industry functions.

In the two tables we see three different categories of countries: (1) large industrialized countries that have large domestic mineral resources; (2) countries with few or no mineral resources; and (3) countries with large resources but relatively small populations or a poorly developed industrial base. The first type of country, for which we could cite as examples Russia, USA and China, are near the top of the list of both producers and consumers; they produce from their own domestic sources a large proportion of the metals that that consume.

In the second category we find countries like Japan and Germany, which possess very few domestic ore deposits but have abundant and active industry that consumes large amounts of raw materials These countries are major importers of ores and/or refined metals. Finally, the countries in the last category – those countries with large resources but small populations or underdeveloped industry – are the major exporters of minerals (Table 2.7). Examples include Australia, South Africa, Chile, Brazil and Jamaica. There are, of course, many exceptions to these general observations. The USA, for example, produces a wide range of mineral products but contains very few large deposits of Ni and Cr. Domestic resources of these metals, which are essential for steel production, are totally inadequate for its needs. The USA is therefore a major importer of Ni and Cr. The incredible

Table 2.7 Major ore-importing and ore-exporting countries from http://minerals.usgs.gov/minerals/pubs/

	Country	Value ($)
(a) Importers		
1.	China	85,280,550
2.	Japan	28,365,440
3.	Germany	9,307,674
4.	Korea	6,623,871
5.	India	5,250,223
6.	United Kingdom	4,679,500
7.	USA	4,487,631
8.	Belgium	3,183,008
9.	Netherlands	3,081,213
10.	Italy	2,912,043
11.	Finland	2,896,519
12.	Canada	2,775,180
13.	France	2,630,696
14.	Russia	2,307,253
15.	Spain	2,217,288
(b) Exporters		
1.	Australia	34,546,550
2.	Brazil	18,726,620
3.	Chile	14,888,160
4.	Peru	7,273,738
5.	South Africa	7,268,294
6.	India	6,519,472
7.	USA	6,487,638
8.	Canada	6,053,128
9.	Indonesia	4,295,629
10.	Sweden	2,628,527
11.	Kazakhstan	2,412,308
12.	Russia	2,374,813
13.	Ukraine	2,153,611
14.	Iran	1,579,345
15.	Congo	1,555,942

Exercise 2.1. Development of a Platinum Deposit in Greenland

The retreat of inland ice in many parts of Greenland has allowed mineral exploration companies to search for new deposits in areas that previously were covered by ice. One of the major targets are deposits of the platinum-group elements. As will be discussed in the following chapter, these deposits are found in layered mafic-ultramafic intrusions, particularly, but not uniquely, those in Precambrian areas.

In this exercise, we ask you to:

(a) Use your knowledge of the geological and tectonic make-up of Greenland to suggest likely areas where exploration could be carried out. (Geological maps and other information is readily available on the internet; e.g. http://www.geus.dk/program-areas/raw-materials-greenl-map/greenland/gr-map/kost_1-uk.htm).

(continued)

Table 2.8 List of the major producers of a selection of metals and ores

	Aluminium	Tonnes x 1000	Bauxite	Tonnes x 1000	Steel	Tonnes x 10^6	Iron Ore	Tonnes x 10^6	Copper	Tonnes x 1000	Gold	Tonnes	Platinum	Tonnes
1	China	16,800	Australia	70,000	China	630	China	900	Chile	5520	China	345	South Africa	138
2	Russia	3,850	China	40,000	Japan	110	Australia	420	Peru	1285	Australia	255	Russia	24
3	Canada	2,920	Brazil	32,100	United States	90	Brazil	370	China	1150	United States	230	Zimbabwe	8.8
4	Australia	1,950	India	18,000	India	67	India	260	United States	1120	South Africa	190	Canada	5.5
5	United States	1,720	Guinea	17,400	Russia	66	Russia	100	Australia	900	Russia	190	United States	3.5
6	Brazil	1,550	Jamaica	9,200	South Korea	56	Ukraine	72	Indonesia	840	Peru	170	Colombia	1.0
7	India	1,400	Kazakhstan	5,300	Germany	44	South Africa	55	Zambia	770	Indonesia	120		
8	United Arab Emirates	1,400	Russia	4,700	Brazil	33	United States	49	Russia	750	Ghana	100		
9	Norway	800	Suriname	3,100	Ukraine	31	Canada	35	Canada	480	Canada	90		
10	South Africa	800	Venezuela	2,500	France	16	Iran	33	Poland	430	Uzbekistan	90		

Source U.S. Department of the Interior I U.S Geological Survey. URL: http://minerals.usgs.gov/minerals/pubs/commodity/copper/index.html. Figures for the year 2010

(b) Discuss the economic, political and environmental aspects of the development of a large platinum deposit in the region. In this discussion you should take into account the geographical position of possible deposits, the climate and other factors that will influence the development of a mine; the distance to likely markets and global trade in the metals; present and future uses of the metals; and finally the political issues – is Greenland a potentially stable supplier of mineral products and how does it compare with other sources.

Elements of a response:

(a) Inspection of the legend of the geological map reveals the presence of mafic intrusions of various ages. These intrusion might contain Ni sulfide deposits like those in Archean komatiites of Australia, or Cr and PGE deposits like those of the Bushveld Complex in South Africa, or magmaic sulfide deposits intrusions related to large magmatic provinces like those in Russia. Descriptions of all these deposits are given in Chap. 3. An exploration geologist would use the map to find areas where such deposits have been located.

(b) When considering the mining of a deposit, factors such as the climate (more extreme in the north and on the east coast); the distance from the coast and means of transporting ore to a site where it could be shipped to future customers; the site of a refinery and possibly a smelter (in Greenland or elsewhere – Outukumpu in Finland is a possibility); measures to be taken to assure that any future mining is conducted in an environmentally correct manner; and finally the relations with the governmental authorities (authority to explore and eventually to mine a deposit, royalties to be paid, hiring of local workers and companies, and so on).

industrial expansion in China has multiplied its need for a wide range of metals and even its large domestic resources cannot meet these requirements. China is therefore a major importer of a wide range of minerals. On the other hand, China contains large deposits of tungsten and produces more than it needs; it is therefore an exporter of this metal. Australia, a country with abundant resources of almost all types of mineral, is a leading exporter of all these minerals, but it lacks major deposits of Cr and is an importer of this metal.

Further complications arise when one distinguishes production of refined metals from unrefined ores. In some cases, ore is exported in its unrefined form, more or less as it is mined. This is the case for some iron ores that are shipped directly from mines in Australia or Brazil to refineries in Japan or China. At the other extreme we have gold or diamond, which almost always are separated from the gangue and refined at the sites where they is mined, and only the pure metal or uncut gemstone is transported out. In most other cases, ores or refined products of variable degrees of purity are exported. Consider, for example, the several major steps in the processing of Zn ore: during mining, the first step, an effort is made to extract only material rich in the ore mineral (sphalerite, ZnS); in the second, the sphalerite is separated in a refinery from the gangue minerals; and in the third, Zn is separated from the sulfide in a smelter. At each step the price of the product increases, commonly by several orders of magnitude. Viewed in this way, it would seem obvious that mineral-producing countries should build refineries and smelters so that they can export the much more valuable end products and not the raw ore; but, as explained in Box 2.3, the situation is not that straightforward.

Box 2.3 Debate About the Politics of Exportation: Raw Materials or Finished Products?

A major dilemma confronts all exporters of mineral products. Should they export unrefined ore or the refined, pure metal or mineral? At first glance the answer seems obvious because the value of the refined product is many times that of the raw ore, and by exporting the latter, the country will earn far more. In addition, the construction and operation of refineries and smelters generates employment, industrial infrastructure and domestic expertise that also are of great benefit to the exporting country.

The counterarguments come from the cost of constructing and operating the refineries and smelters. In order for the operation of these factories to be economically sound, the operation must be of a certain size; if the ore deposits are small, the construction of even a refinery, not to mention a smelter, is commonly not viable. In addition, a large investment is needed to construct the factory and many countries do not have funds for investment; they have to be borrowed and interest has to be paid.

Consider the following example. A deposit of nickel containing 10 mt of ore with 2% Ni is found in Zimberia, a small country in central Africa. The deposit contains 200,000 t of Ni of which about 70% can be extracted, given a total of 140,000 t. If sold as refined metal, the nickel is worth a total of about two billion euros at current metals prices of about 15,000 euros per tonne. In the future the Ni price and the amount earned are likely to increase. The raw ore is sold at about 70 euros per tonne, giving a total value of the unrefined ore of only 140 million euros, about 7% of the value of the refined metal (These figures do not take into account the cost of mining, refining and exporting the ore or metal, which we will ignore in this exercise). The expected lifetime of the deposit (i.e. the time before it is completely mined out) is 20 years. Other deposits, may, however, be found in the region in the future.

The total cost of building the refinery and smelter needed to purify the metal is about three billion euros. Zimberia does not have the money itself and would have to borrow on the international market. The total cost of borrowing this money for a period of 20 years is about the same as the capital amount. Discuss whether it is worthwhile for Zimberia to construct a Ni refinery so that the country can export refined metal and not the raw ore.

In Chap. 5 we discuss a curious situation. Because of the large amounts of energy needed to refine aluminium, it is worthwhile shipping bauxite from countries like Jamaica or Australia, where it is mined, clear across the globe to Iceland, where it is refined. This exercise, which no doubt causes nightmares for ecologists worried about shipping beans from Kenya to Germany, is made possible because of the availability of abundant, cheap, greenhouse-gas-free sources of hydro and geothermal energy in that small island in the middle of the North Atlantic ocean. Also discussed in Chap. 5 are the problems that are associated with the production of ore in some third-world countries. The case in point are the massive and rich copper deposits of central Africa, which should have provided wealth for the extremely poor countries of the region but never have because of exploitation by colonising countries at first, and the corruption and inefficiency of the governments of newly independent countries thereafter.

2.5   World Trade in Mineral Resources

Patterns of world trade in minerals are summarized in Fig. 2.2. The arrows, which show the direction of trade, link copper-producing countries like Chile to large industrialized countries or regions like the USA, Japan, China and Europe. The geographic position of the producer and consumer influences the direction of trade: most Chilean copper is shipped across the Pacific to USA, Japan and China and not around Cape Horn to Europe (large ore carriers cannot pass through the Panama Canal). But other factors enter into the equation. Australia has a distinct geographical advantage when selling iron ore to China and Japan, and would be capable of supplying most of the needs of these countries, but these countries also import ore from Brazil. This ore must be shipped around Cape Horn, but the consuming countries are willing to pay the extra transport costs so as to give them some leverage when negotiating the price they pay for the ore. If Australian producers had a monopoly on the market, they could charge higher prices.

Also seen in the diagram is the precarious nature of the sources of some commodities. At present the major producers of the platinum group metals, which are essential for modern industry, particularly for the fabrication of catalytic converters in cars, are located in three countries. One of these, Canada, will remain a stable, reliable supplier for the foreseeable future. The second is Russia, a country whose reputation as a supplier of natural resources has been compromised by its manipulation of gas supplies to Europe. The third, and by far the major supplier, is South Africa, thanks to the enormous resources in the Bushveld Complex (Chap. 3). As long as the democratically elected, stable and still relatively honest ANC government survives, and as long as it continues its market-oriented policies, the country will remain a reliable supplier of the platinum-group metals. But if government policies in that country were to change, or if social peace were compromised,

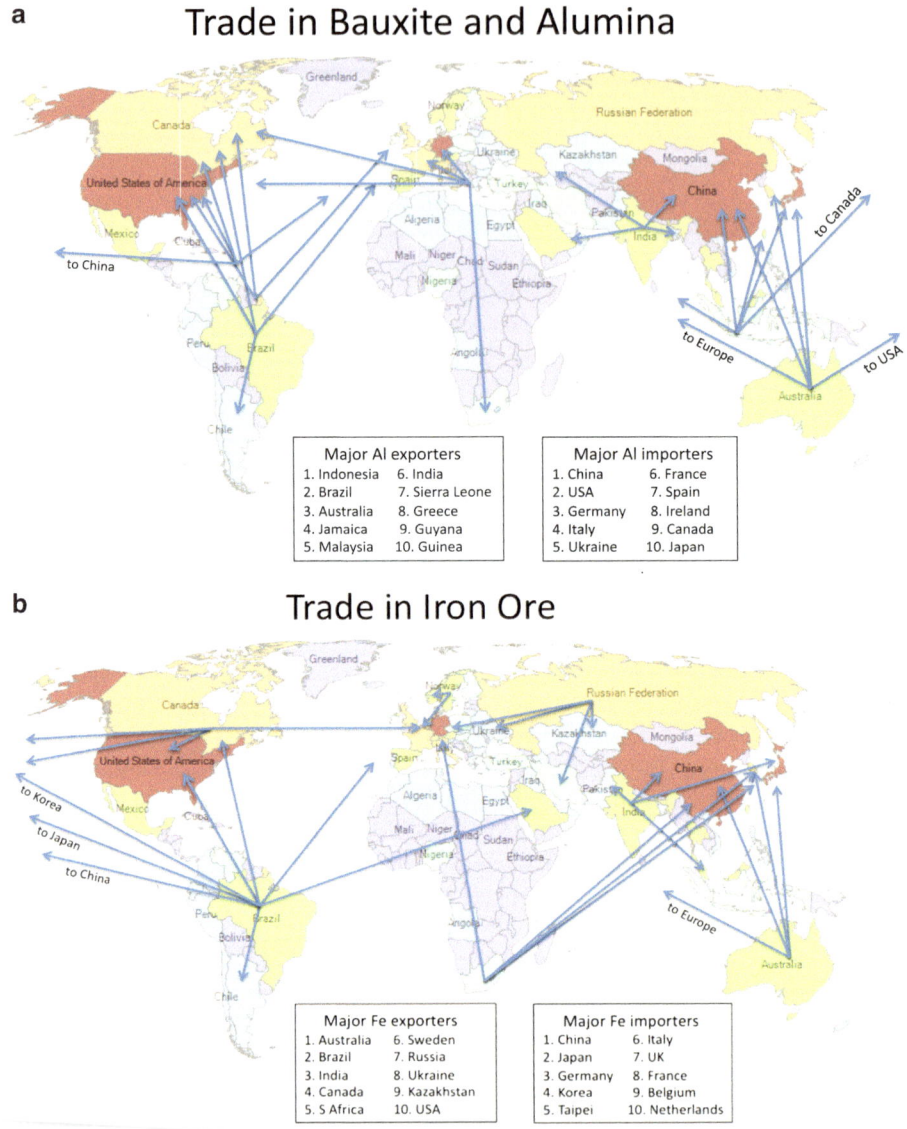

a

Trade in Bauxite and Alumina

Major Al exporters	
1. Indonesia	6. India
2. Brazil	7. Sierra Leone
3. Australia	8. Greece
4. Jamaica	9. Guyana
5. Malaysia	10. Guinea

Major Al importers	
1. China	6. France
2. USA	7. Spain
3. Germany	8. Ireland
4. Italy	9. Canada
5. Ukraine	10. Japan

b

Trade in Iron Ore

Major Fe exporters	
1. Australia	6. Sweden
2. Brazil	7. Russia
3. India	8. Ukraine
4. Canada	9. Kazakhstan
5. S Africa	10. USA

Major Fe importers	
1. China	6. Italy
2. Japan	7. UK
3. Germany	8. France
4. Korea	9. Belgium
5. Taipei	10. Netherlands

Fig. 2.2 (continued)

then there would be a global crisis of supply of these metals. It is for partly these reasons that platinum has the highest criticality index of all metals evaluated by the US Geological Survey (Box 1.1). What would be the consequences a global crisis? In the short term, the metal price would skyrocket and the cost of producing cars and other products that use the metals would increase; in some cases production might stop, if insufficient supplies could not be found. On the longer term, however,

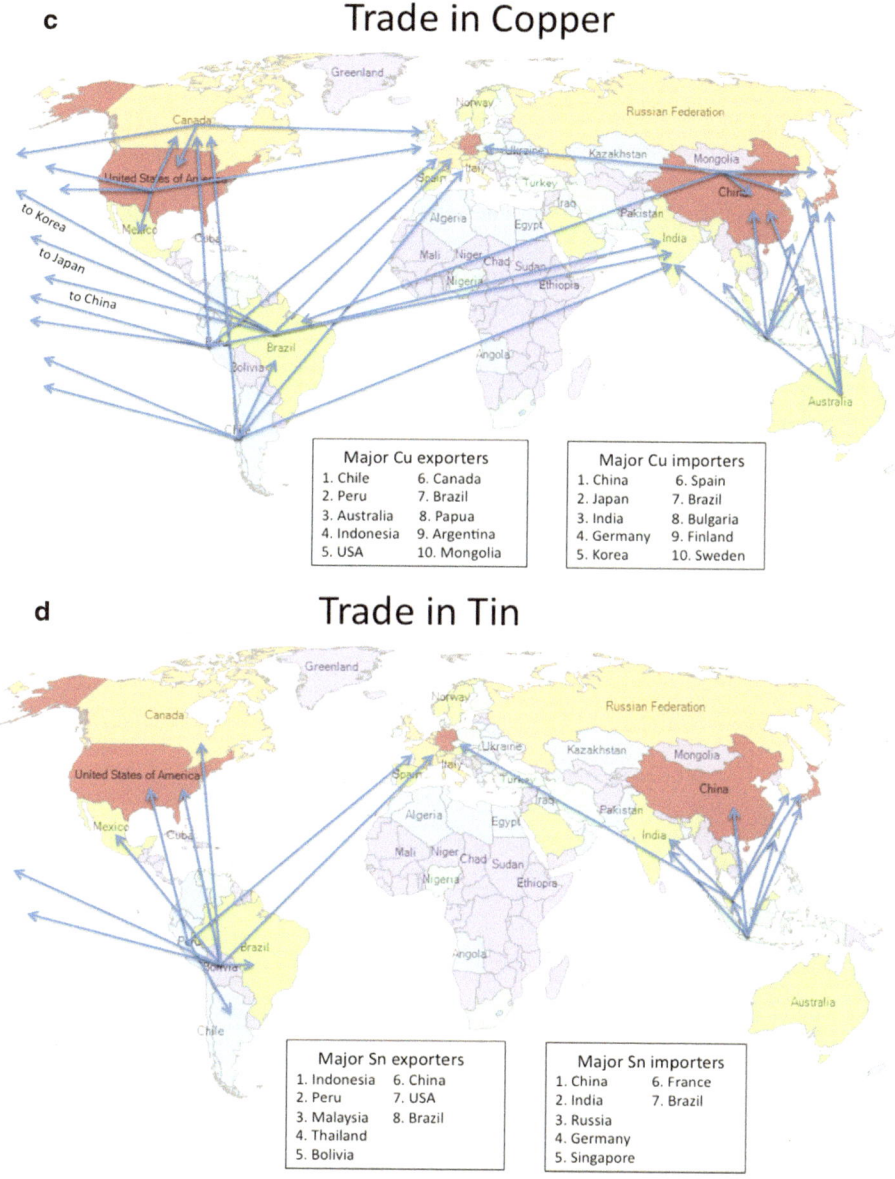

Fig. 2.2 Global trade in minerals. (**a**) Trade in bauxite and alumina. (**b**) Trade in iron core. (**c**) Trade in copper. (**d**) Trade in tin

the higher metal price would, on one hand, allow companies to open mines in hitherto marginal deposits or conduct exploration to find new deposits; and on the other hand, replacements would be sought and found for the industrial uses of the metals. These issues will be revisited in Chap. 6.

2.6   General Sources

British Geological Survey, World Mineral Production 2005–2009. http://www.bgs.ac.uk/mineralsuk/statistics/worldStatistics.html.

Canadian Geological Survey, Geological Data Repository. http://gdr.nrcan.gc.ca/minres/index_e.php

USGS Mineral Resources Web sites. http://minerals.usgs.gov/ http://minerals.usgs.gov/minerals/

Trade Map – Trade statistics for international mineral trade. http://www.trademap.org/Country_SelProduct_Map.aspx

References

Evans AM (1993) Ore geology and industrial minerals, an introduction. Blackwell Science, Oxford, UK. ISBN 0-632-02953-6

Lindgren W (1933) Mineral deposits, 4th edn. McGraw-Hill, New York

Chapter 3
Magmatic Ore Deposits

3.1 Introduction

A magmatic ore deposit is an accumulation of magmatic minerals. Some of these minerals are extremely rare and almost never encountered in common rocks, an example being alloys of the platinum metals; other minerals, such as magnetite, are common and can be seen in many thin sections. They form an ore deposit when they accumulate in large amounts and at unusually high concentrations. The question is how these concentrations come about.

3.2 Chromite Deposits of the Bushveld Complex

To illustrate the ore-forming process, we will take as our first example the deposits of chromite in the Bushveld Complex of South Africa. Chromite, a Cr-Fe oxide, is the only ore of the metal chromium. A brief description of the complex, emphasizing its economic importance, is given in Box 4.1. Bushveld contains the type examples of ore deposits in a large layered intrusion, but this is not the sole type of chromite deposit; others are found in ophiolites, particularly in the Urals, Turkey, Greece and India. In addition to their great economic value, the deposits of the Bushveld were chosen because they illustrate several fundamental processes governing ore formation in a magmatic setting.

Figure 3.1, a photo of chromite veins from the Dwars River location in the eastern side of the complex, clearly shows some important features of a chromite deposit. The mineral occurs in layers that may reach a metre or more in thickness, which alternate with layers composed of other magmatic minerals. The rock is a cumulate, having formed by the transport of magmatic minerals to the floor of the Bushveld magma chamber. The Bushveld Complex itself (Fig. 3.2) is a vast, roughly funnel-shaped differentiated intrusion composed of the Lower Zone which consists of alternating layers of cumulus mafic minerals such as olivine and

N. Arndt and C. Ganino, *Metals and Society: an Introduction to Economic Geology*,
DOI 10.1007/978-3-642-22996-1_3, © Springer-Verlag Berlin Heidelberg 2012

Fig. 3.1 Chromite veins at
Dwars River, Bushveld
Complex. The black bands,
whose average thickness is
about 10–20 cm, are chromite
cumulates; the white bands
are plagioclase cumulates

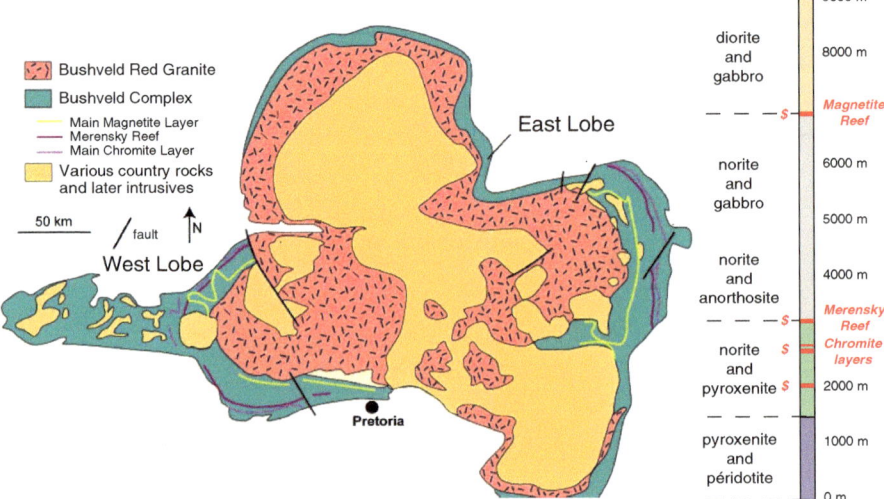

Fig. 3.2 The best examples of magmatic ore deposits: magnetite, chromite and platinum group elements in the Bushveld Complex, South Africa

pyroxene, the mafic Main Zone of pyroxene and plagioclase cumulates, and the more evolved Upper Zone of diorite or gabbro. Intervening between Lower and Main Zone is the Critical Zone, which contains the chromite and PGE deposits. More detailed descriptions of the complex and its ores are found in Cawthorn et al. (1996, 2005).

Chromite, the ore mineral, is present throughout the lower ultramafic part of the intrusion, but normally its concentration is less than 1–2%. At these levels the rock is not ore. Only at specific levels in the Complex, in olivine ± orthopyroxene cumulates in the upper part of the Critical Zone (Fig. 3.2) is chromite present in sufficient quantities and at sufficient concentrations, to constitute an ore deposit. The chromite cumulate layers consist of close to 100% of the ore mineral, and typical ore grades are around 25–35% Cr. The concentration of the same element in

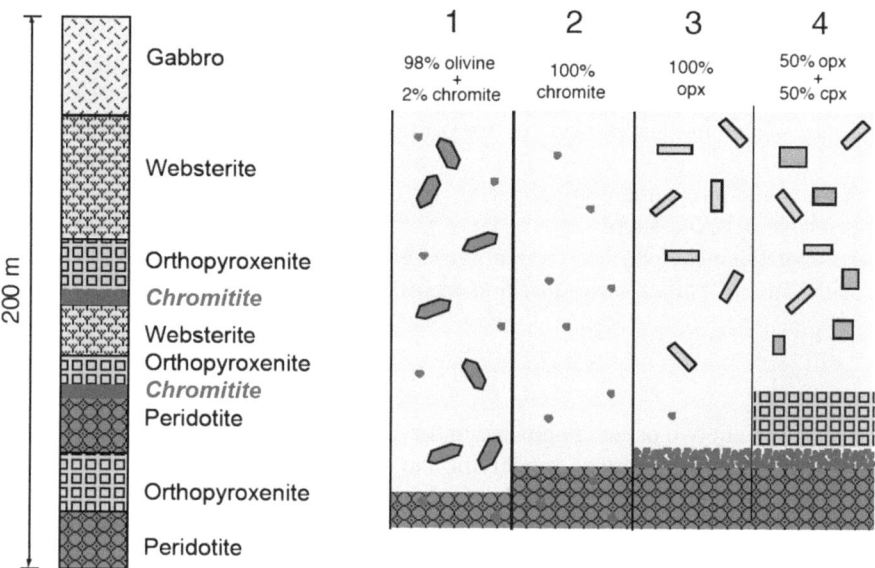

Fig. 3.3 Formation of chromite veins in the Muskox Intrusion, Canada

the continental crust is only a few tens of ppm, but this figure is not relevant because chrome deposits are found in ultramafic rocks in which concentrations are much higher. But even taking a typical level for ultramafic intrusions of 1,000 ppm means that the enrichment factor; i.e. the ratio of the ore grade over the background level, is 250–350. In other words, to form the ore deposit, the concentration of Cr had to be increased by several hundred percent. What geological process could have produced this degree of enrichment?

Box 3.1 Deposits of the Bushveld Complex
The Bushveld complex in South Africa (Fig. 3.2) is the world's largest mafic-ultramafic intrusion. As shown in the figure it has a complex form and consists of a collection of intrusions that are roughly circular in plan in plan view; in section it probably comprises a series of amalgamated funnel-shaped (lopolithic) magmatic bodies. The dip around the periphery is persistently towards the interior and thus the base of the intrusion is exposed along the margin. The stratigraphic section shows that the rock-types change from ultramafic at the base to intermediate at the top. The lowermost rocks are olivine and pyroxene cumulates; these pass upwards through pyroxene pla-gioclase cumulate to the diorites that occupy the upper third of the complex.

Three types of ore deposit are found in the complex. (1) Chromite deposits, such as the layers illustrated in the photo, are located in the ultramafic

(continued)

cumulates. These deposits are mined for chrome and in some cases for platinum group elements as well. (2) Platinum group elements are also mined in the famous Merensky Reef, which is located at the top of the Critical Zone, where the rocks change from ultramafic to mafic in composition. (3) Finally, magnetite veins, which are mined for the vanadium, are hosted in the mafic rocks in the upper part of the complex.

These deposits supply a very large proportion of the global demand for these metals. The chrome production is about 50% of the global total and platinum and palladium production from the complex represents 72% and 34%, respectively, of annual global production.

Neil Irvine, in two papers published in the 1970s (Irvine 1975, 1977), developed an interesting and important model for the origin of chromite deposits in the Bushveld Complex and in other layered intrusions. His diagram, modified as Fig. 3.3, shows that under normal conditions 1–2% chromite crystallizes together with olivine. Mafic-ultramafic liquids, such as the parental magma of the Bushveld Complex, plot in the field of olivine. Such a liquid initially crystallizes olivine, whose removal drives the residual liquid composition to the cotectic, at which point chromite starts to crystallize. Chromite and olivine crystallize together, in the proportion given by the intersection of a tangent to the cotectic with the olivine-chromite side of the diagram. The proportion of chromite ranges from 1.8% to 1.4%, the amount observed in normal olivine cumulates of the complex. To produce a layer of almost pure chromite requires suppression of the accumulation of olivine and other silicate minerals. Irvine suggested two ways this might happen, both illustrated in the phase diagrams of Fig. 3.4.

In this simple diagram, a strongly curved cotectic separates the primary phase fields of olivine and chromite. For chromite to crystallize alone, the composition of the liquid must be driven off the cotectic and into the chromite field. Irvine (1975) proposed that one way that this might happen is when the magma becomes contaminated with granitoid country rock like that which makes up most of the Archean continental crust into which the Bushveld magma intruded. Granite plots at the SiO_2 apex in the diagram. A hybrid, contaminated magma plots on a line between the magma composition and the apex, at the point labelled D in the chromite field. This magma crystallizes chromite alone until its composition regains the cotectic. The interval in which chromite crystallizes seems small, but in a large intrusion like the Bushveld, the chromite that crystallizes during the interval is enough to form a layer of appreciable thickness monomineralic chromite cumulate.

The second process depends on the strongly curved shape of the cotectic. Irvine (1977) argued that because of this shape, when an evolved liquid residing in the chamber mixes with a more primitive liquid that enters the chamber, the hybrid liquid plots within the chromite field. Like the contaminated liquid, the hybrid

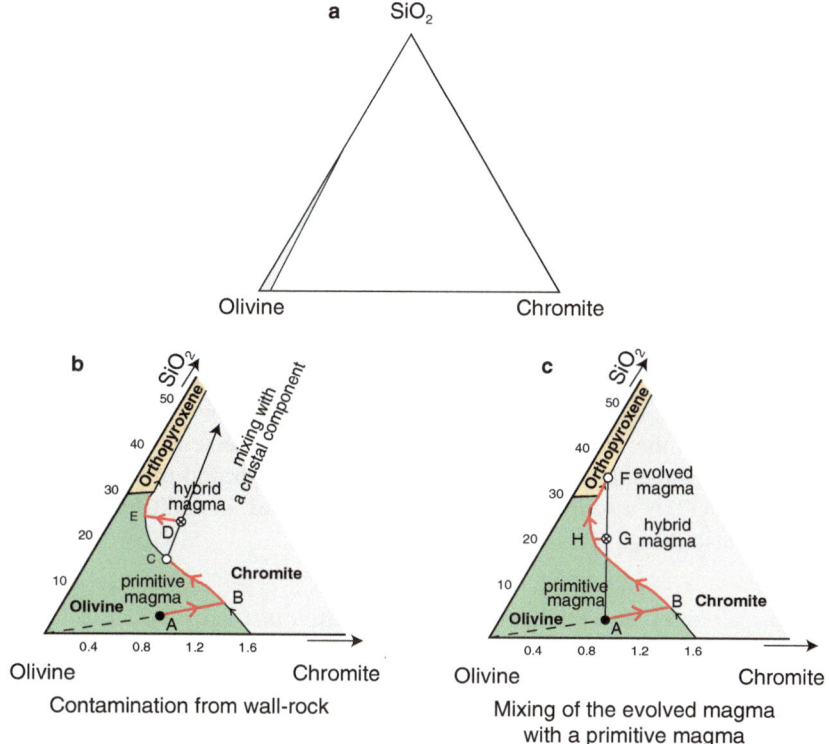

Fig. 3.4 Irvine's mechanisms explaining how chromite can crystallize alone. (**a**) sketch of phase diagram showing position of diagrams b and c, (**b**) Contamination model, (**c**) magma mixing model

liquid crystallizes chromite alone. Most probably both processes – contamination and magma mixing – operate together to produce the chromite deposits of the Bushveld Complex. Obviously the process is not that simple; in particular, it is very difficult to understand how the chromite crystals accumulated and how they were extracted from the enormous volume of magma needed to yield metre-thick layers of chromite.

Nonetheless, the formation of chromite deposits illustrates an important principle: the ore mineral, in this case chromite, is a normal constituent of many ultramafic intrusions and it forms through normal magmatic processes. Under ordinary circumstances it is present in low concentrations: normal magmatic rocks are not ores. For a deposit to form, the normal geological process must be perturbed so that the ore mineral accumulates in far higher concentrations. In the case of the chromite deposits, contamination or magma mixing are among the perturbing processes. As we will see in subsequent chapters, a large range of special circumstances modifies other geological processes like sedimentation or the circulation of hydrothermal fluids so as to create the unusual concentrations of minerals that constitute an ore.

3.3 Magnetite and Platinum Group Element Deposits of the Bushveld Complex

The Bushveld Complex contains two other important types of deposit. Layers of magnetite, mined for their vanadium contents, occur in the upper part of the intrusion. These magmatic deposits probably formed in a manner similar to the chromite deposits. The other type, economically far more important, are deposits of the platinum group elements (PGE). The Bushveld Complex contains about 60% of global reserves of these increasing valuable metals, mainly at two specific horizons in the lower part of the intrusion. The upper layer is the famous Merensky Reef, a thin (1–10 m) layer of pegmatoid pyroxenite located at the top of the Critical Zone (Fig. 3.2). Seemingly low, but economically viable, concentrations of PGE (5–500 ppm) are associated with minor sulfides mainly towards the base of the Reef. Because of the high cost of the platinum group elements, even these low concentrations of metals can be mined at a profit. The second major mineralized layer, called UG2, is a series of thick chromite reefs that, in addition to high PGE content, are also mined for their Cr contents.

Box 3.2 Competing Theories for the Formation of the PGE Deposits

There no consensus about the origin of the PGE deposits. One school, championed by Tony Naldrett and Ian Campbell (Campbell et al. 1983; Naldrett 1989) and Grant Cawthorn and colleagues (Cawthorn et al. 2005), argues that these deposits formed through magmatic processes; the opposing school, led by Alan Boudreau, Chris Ballhouse, Ed Mathez and others (Ballhaus and Stumpfl 1986; Boudreau 1995; Mathez 1995) ascribes an important role to the migration of volatile-rich fluids.

Campbell and Naldrett proposed that a plume of primitive magmatic liquid was injected into the base of the chamber and then mixed with evolved liquid to produce a hybrid magma that became saturated in sulfide. Small droplets of magmatic sulfide segregated from the silicate liquid and these attracted the chalcophile (sulfur-loving) PGE. The droplets of PGE-enriched sulfide then settled to the floor of the intrusion to slowly build up the ore-bearing horizon. Geologists of the opposing school propose that volatile-rich fluids migrated up through the cumulus pile, leaching out the PGE from the cumulus minerals then redepositing them at favourable horizons. The various models are illustrated in Fig. 3.5.

What do you make of these competing hypotheses? Is it not surprising that the origin of some of the world's greatest ore bodies, which are well exposed because of good outcrop and extensive mining operations, and which have been intensively studied for almost a century, is still so poorly understood? Why is it that two groups of highly respected, highly experienced geologists have developed such different models? Which model do you think is the more plausible (to answer this question will require that you read some of the abundant literature available on the subject)? What type of research could be done to (help) resolve the issue?

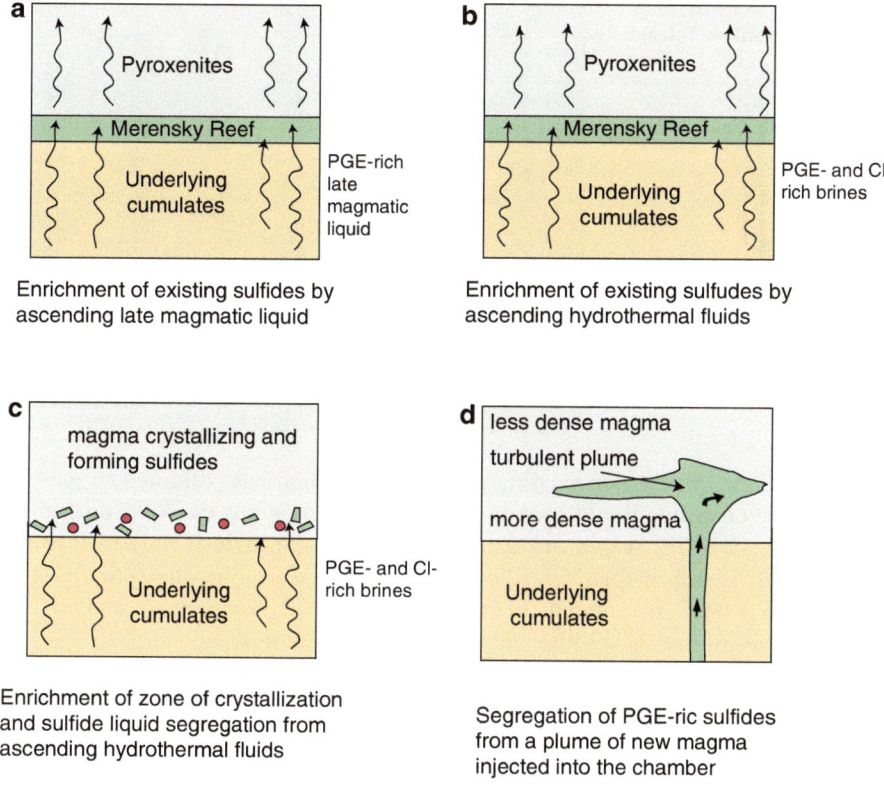

Fig. 3.5 Models for the formation of the Merensky Reef (From Naldrett (2004))

Platinum-group elements are recovered from other large layered mafic-ultra-mafic intrusions like the Stillwater in the USA and the Great Dyke in Zimbabwe, and these constitute the major source of these metals. The same metals are also recovered, though in differing proportions, as a valuable by-product of mining magmatic sulfide deposits, our third example of a magmatic ore deposit, and in placer deposits, which are discussed briefly in Chap. 5.

3.4 Magmatic Sulfide Deposits

When a mafic magma cools it crystallizes. A series of solid phases appear, typically olivine, pyroxene, feldspar and oxides, which become the constituents of mafic and ultramafic magmatic rocks. Under some circumstances second liquid separates, in rare cases another silicate liquid, which is immiscible with the first silicate liquid, or, if the sulfur content of the liquid is high, an immiscible sulfide liquid. In the remarkable example shown in Fig. 3.6, droplets of sulfide liquid have been frozen in

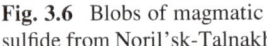
Fig. 3.6 Blobs of magmatic
sulfide from Noril'sk-Talnakh

place, suspended in the surrounding mafic liquid which has solidified to gabbro.
The droplets of sulfide liquid contain elevated concentrations of elements such as
Ni, Cu and the PGE, which are chalcophile or sulfide-loving, and because the
sulfide liquid is denser than the silicate liquid, they tend to settle to the base of
the magma body. If enough of the sulfide liquid segregates, and if the
concentrations of metals are high enough, it becomes an ore deposit.

This process seems straightforward, yet once again we are faced with a problem:
mafic-ultramafic intrusions are known throughout the world, and from petrological
and geochemical data we can infer that their parental magmas contained high
concentrations of Ni, Cu, and the PGE. Inspection of polished thin sections shows
that many of these rocks do indeed contain sulfides that separated as an immiscible
liquid, but in most cases this phase appears only at a very late stage in the
crystallization sequence and only in very low quantities. What was the particularity
of certain magmas that led them to segregate copious amounts of metal-rich
sulfide? Or to put it another way, how was the normal magmatic differentiation
sequence perturbed so as to form an ore deposit?

To answer these questions we will first discuss the processes that govern whether
or not a sulfide liquid separates from a mafic or ultramafic magma, and, equally
important, the concentrations of ore metals in this liquid. Then we consider how the
sulfide segregates to form an ore body, taking as our first example the deposits of
the Kambalda region in Western Australia.

3.4.1   Controls on the Formation of Magmatic Sulfide Liquid

Mafic and ultramafic magmas form by partial melting at great depth in the mantle,
from about 30 to 300 km. Most of the magmas that yield magmatic ore deposits
result from melting in mantle plumes, which are cylinders or more irregular masses
of solid mantle that are hotter than the surrounding ambient mantle and ascend
because of their low density. Only part of the mantle peridotite melts, between 5%

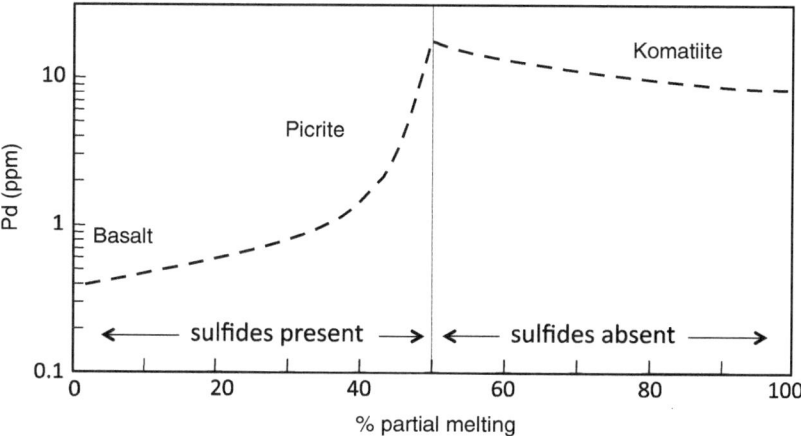

Fig. 3.7 Controls on the chalcophile element contents of mantle-derived melts

and 20% for basalts, 20–30% for picrites and up to 60% for komatiites, and the composition of the magma depends strongly on the types of minerals that remain in the residue. The overall character of the melt is controlled by the silicate phases in the residue of melting while the contents of ore metals like Ni, Cu and the PGE depend on the sulfide. In normal mantle peridotite, sulfur is present as sulfide, which enters to liquid at the start of the melting process and is totally exhausted when the degree of melting exceeds 20–30%. As shown in Fig. 3.7, sulfide is retained in the residue during the formation of low-degree melt like basalt but is exhausted in high-degree melt like komatiite. Nickel, Cu and the PGE are all strongly chalcophile, which means that when sulfide is retained in the residue, it holds back these elements: the resultant magmas contain only low concentrations. High-degree melts, on the other hand, acquire their full component of these metals, and partly for this reason are the most prone to form ore deposits.

When the magma enter the crust and starts crystallize, an immiscible sulfide liquid will separate from the silicate liquid if the concentration of sulfur exceeds the sulfide solubility. The situation can be compared with crystallization of salt from brine. Only if concentration of salt is high enough will the brine become saturated and salt crystallize. But saturation can also be reached if the brine evaporates, which decreases the amount of water and increases the salt concentration in the remaining brine; or the solubility of salt can be decreased by changing the temperature or pressure, or by adding to the solution other components that decrease the solubility. The same principles apply to the separation of a sulfide liquid from a silicate liquid. Experimental studies have shown that the solubility of sulfide depends on external parameters such as temperature and pressure, and on the composition of the melt. Table 3.1 summarizes these factors.

During fractional crystallization of magma, the temperature drops, the Fe content usually varies little and the Si content increases. The process will therefore lead eventually to sulfide saturation and the separation of sulfide liquid. The process can

Table 3.1 Controls on the solubility of sulfide in silicate melts

Factors that increase sulfide solubility
– Increasing temperature
– Increasing Fe content
Factors that decrease sulfide solubility
– Increasing pressure
– Increasing Si content

With increasing oxygen fugacity, the speciation of sulfur changes; at low fO_2, it dissolves as sulfide; at higher fO_2 it is present as sulfate and the solubility is far higher

be accelerated if the magma is contaminated with granitoid country rock, which increases Si and decreases Fe; or it can be caused by the assimilation of sulfide- or sulphate-bearing sediments, which directly increases the S content of the melt.

Mavrogenes and O'Neill (1999) showed that sulfide solubility varies inversely with pressure. This means that magma that formed at high pressure deep in the mantle is capable of dissolving less sulfide than the same magma at lower pressure in the crust. The consequence is that most magmas from deep in the mantle are moderately or highly undersaturated in sulfide when they intrude into the crust. For such a magma to segregate sulfide liquid, either it must crystallize almost completely (in which case the sulfide will be trapped between abundant crystal and cannot accumulate to form an ore deposit) or the system must be perturbed so that the sulfide segregates sooner.

3.4.2   Controls on the Segregation and the Tenor of Magmatic Sulfide Liquid

The value of a sulfide ore deposit varies widely, depending on the concentrations of ore metals in the sulfide phase. Some large accumulations of sulfide contain very low concentrations of Ni, Cu and PGE (they consist essentially of pyrrhotite, $FeS_{(1-x)}$ and pyrite, FeS_2) and they do not constitute an ore deposit. Other deposits contain a high proportion of Ni-rich sulfides like pentlandite $(Ni, Fe)_9S_8$ or better still millerite or nicolite, and they form an ore body even if the amount of sulfide is low.

The ore metals Ni, Cu and the PGE are all chalcophile and have a tendency to partition more or less strongly into the sulfide. Nickel is lithophile as well as chalcophile and in normal ultramafic rocks it is distributed between olivine and sulfide. Copper is moderately chalcophile (the partition coefficient $KD^{sulf\text{-}silicate}_{liquid}$ is about 100), but the PGE are enormously chalcophile, having a KD of 10^{4-5}. This means that any droplet of sulfide will extract most of the Cu and Ni, and effectively all of the PGE, from the surrounding silicate liquid. If the sulfide droplets can then be concentrated efficiently, for example by gravitative settling, then an ore deposit forms.

In practice, other processes intervene. The metal contents of the magma clearly influence composition of the ore: it is evident that magma containing little to no Ni cannot produce a nickel deposit. Ultramafic magma has high Ni but low Cu contents and sulfide ores in ultramafic rocks have high Ni/Cu. More important, however, is the relative proportion of sulfide to silicate liquid, which influences the proportion of the chalcophile elements that are present in the sulfide. The situation is best illustrated by the extremely chalcophile PGE, which, at equilibrium, will be almost entirely contained in the sulfide phase. However, in a static system, the PGE migrate into the sulfide by diffusion, which is inefficient. In the absence of mechanical mixing of the two phases, each sulfide droplet will be surrounded by a zone of silicate liquid that is effectively stripped of PGE. Only if the sulfide is mixed with a large volume of silicate liquid can it realize the high PGE content promised by the high partition coefficients. To describe this process, Campbell and Naldrett (1979) introduced the R-factor, which measures the relative proportion of silicate liquid that interacted with sulfide liquid. If the R-factor is low; i.e. a small volume of silicate liquid is mixed with a large volume of sulfide, the content of PGE and other chalcophile elements of the sulfide is low. This is the case for ore deposits in small intrusions or lava flows, like the Kambalda example discussed below. If, however, a small volume of sulfide can mix with a large volume of silicate liquid, the chalcophile element content of the sulfide is high. This is the case for the magmatic sulfide of the Merensky Reef, which we discussed in the previous section (Campbell et al. 1983).

3.4.3   Kambalda Nickel Sulfide Deposits

We chose not to start with the very largest and richest Ni-Cu sulfide deposits, which are found in intrusive rocks of various types and origins, but will first investigate the ores at Kambalda, one of very few deposits that occur a volcanic setting. Kambalda is located in hot, dry savannah of the Western Australian outback, in the Archean (2.7 Ga) Yilgarn Craton. The geological make-up of this region, summarized from Marston et al. (1981), Lesher (1989) and Lesher and Keays (2002), is shown in Fig. 3.8: a series of ultramafic lava flows (komatiites), is underlain by tholeiitic basalts and overlain by magnesian basalts. The complete sequence is exposed in a small structural dome. Figure 3.9 is a schematic cross section through the lava pile. The ore deposits are mainly restricted to the lowermost komatiite flow and they are localized at the base of this flow, within troughs in the underlying basaltic sequence. Away from the ore deposit, thin bands of sulfide-rich cherty sediment intervene between basalt and komatiite but these sediments are missing within the troughs that contain the ore.

The ores themselves have features that need to be catalogued because they provide important clues as to the ore-forming process. Strictly speaking, the ore minerals should be described as Fe-Ni-Cu-PGE sulfides because they contain all of these metals. The main ore minerals are pentlandite $(Fe,Ni)_9S_8$ and chalcopyrite $CuFeS_4$ which coexist with the barren Fe sulfide pyrrhotite $(Fe_{(1-x)}S)$. In many ore sections, a layer of massive 100% sulfide lines the base of the komatiite flow and is overlain by "net-textured" ore, in which serpentinized olivine grains are enclosed in a sulfide matrix, and in turn by (serpentinized) olivine cumulate containing disseminated sulfides. Veins and lenses of Cu-rich sulfides penetrate into the floor

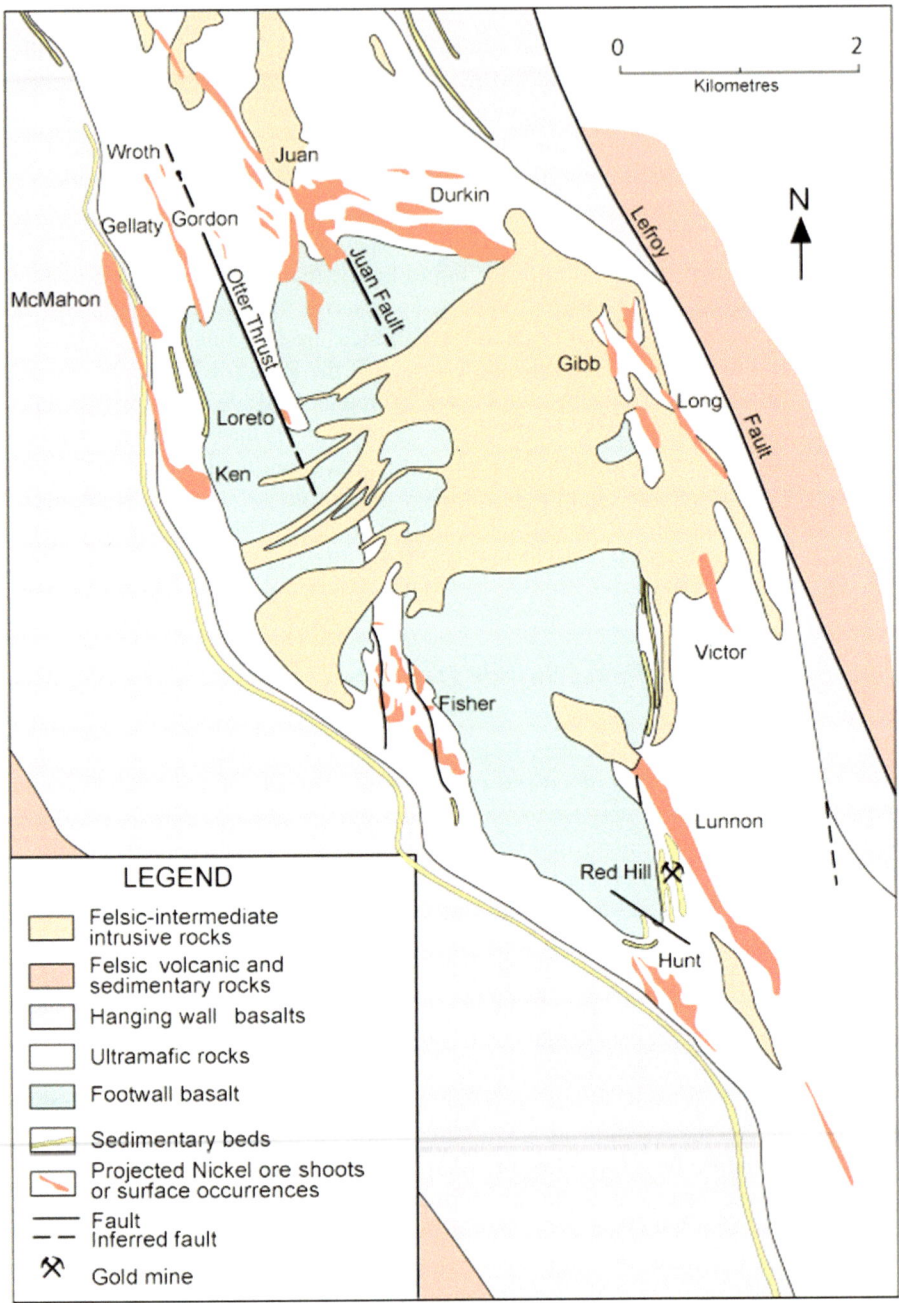

Fig. 3.8 Geological map of the Kambalda region, Australia from Marston et al. (1981) and Lesher and Keays (2002)

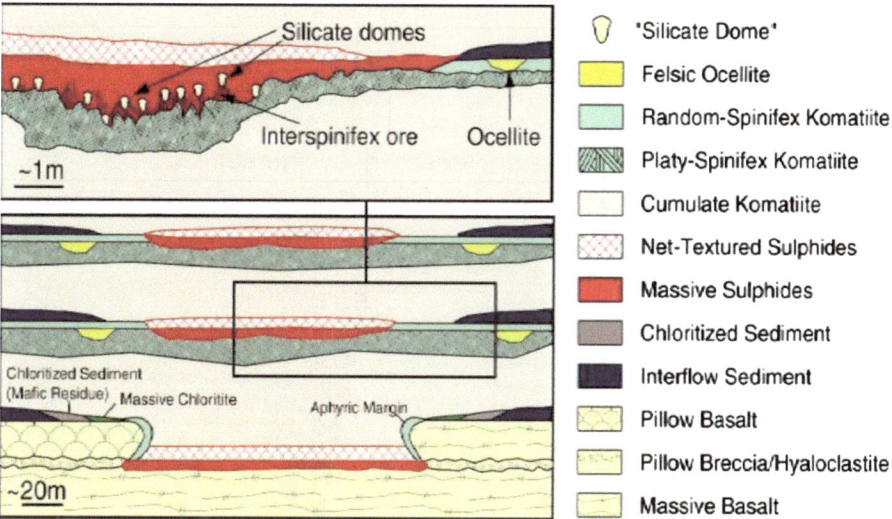

Fig. 3.9 Schematic cross-section through a typical Kambalda ore shoot, showing distributions of interflow metasediments, interspinifex ores, and felsic ocellites (After Groves et al. (1986), Frost and Groves (1989))

rock and in places invade the upper part of the komatiite flow. The composition of the disseminated sulfide corresponds to "monosulfide solid solution", a term that refers to the immiscible sulfide liquid that separates from the silicate liquid. The massive sulfide liquid undergoes fractional crystallization (just like the silicate liquid fractionally crystallizes). The first solid sulfide that crystallizes is relatively rich in Ni and Fe and this material remains as a "cumulate" layer at the base of the flow while the late-solidifying Cu- and PGE-rich sulfide liquid may leak out to form veins in surrounding rocks.

Komatiite lava flows occur throughout the 1,500 km long Yilgarn Craton, but ore deposits are known in only some of them. And komatiites are common in the much larger Abitibi belt in Canada, but there the ore deposits are small and rare. With this background we are in the position to ask a number of questions:

- Why do the sulfide deposits occur preferentially in the lower part of the lowermost komatiite flow?
- Why are sedimentary rocks present between each komatiite flow, except in the troughs that contain the ore deposits?
- Why do Ni deposits form in komatiites and not in basaltic lava flows?
- Why are Ni sulfide deposits more common in the Yilgarn than in the Abitibi belt?
- What distinguishes the ore-bearing Kambalda komatiites from barren komatiites?

Table 3.2 compares the chemical and physical properties of komatiitic and basaltic magmas. The ultramafic magma has much a higher MgO content and a lower SiO_2 content than the basaltic magma, which means that it erupts at much

Table 3.2 Comparison of the physical properties of komatiites and basalts

	Komatiite	Basalt
MgO (wt.%)	30	8
Temperature (°C)	1600	1200
Viscosity (poises)	5	100
Reynolds number (turbulent flow when Re > 500)	10^4 to 10^5	500

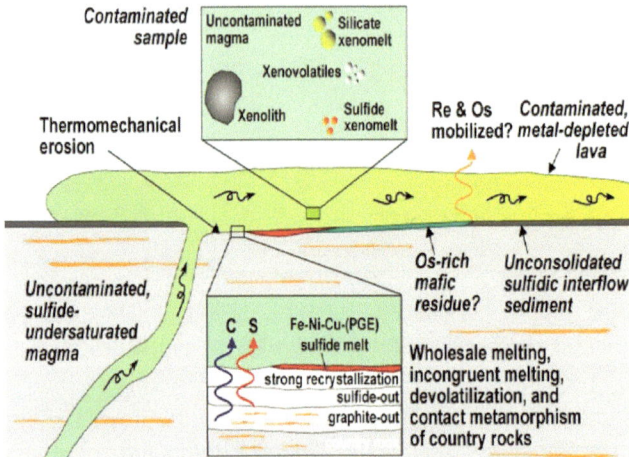

Fig. 3.10 The ore-forming process at Kambalda

higher temperature (up to 1,600°C), and has a much lower viscosity, than the basalt. The Reynold's Number is a fluid-dynamic parameter that describes whether the flowage of a liquid is linear or turbulent. A threshold of 500 separates the two flow regimes. The value for a 10 m thick komatiite flow is around 10^5, well above the threshold. The komatiite lava therefore flows turbulently and the heat from this ultra high-temperature lava is transferred directly to the floor rocks (Fig. 3.10). The consequence is that the floor rocks melt and are assimilated into the komatiite lava. At Kambalda, the floor rocks are sulfide (pyrite)-rich cherty sediments which, when assimilated into the komatiite, change the composition of the magma. This contamination simultaneously boosts the S content and decreases sulfide solubility and this leads to the segregation of immiscible sulfide liquid. Because the komatiite has a high Ni content, and because Ni is a chalcophile element, the sulfide becomes rich in Ni and other chalcophile elements like Cu and the PGE; because the sulfide is denser than the silicate liquid, it settles to the base of the flow to form the ore deposit. The process is illustrated in Fig. 3.10.

Basaltic liquids are cooler and more viscous than komatiites and their low Reynold's number means that their flowage was laminar. For this reason they are rarely capable of assimilating their floor rocks. Extra sulfur cannot readily be incorporated from external sources and a separate sulfide liquid forms only at late stage of crystallization when abundant crystals prevent its segregation. In addition, because the basalt is relatively poor in nickel and PGE (because sulfide is retained in the mantle source), the sulfide contains only low tenors of these elements. These are the main reasons why deposits of this type are found in komatiitic and not basaltic flows.

Why are the deposits particularly common in the Kambalda region? There are two contributing factors. The first is the presence of the sulfur-rich sediments which provide the sulfur; the second is the nature of the komatiites themselves. The lava flows that host the ore deposits of the Kambalda dome are unusually thick (up to 100 m) and they consist mainly of cumulates of Fo-rich olivine. These characteristics correspond to those of rocks precipitated from a relatively primitive komatiite liquid, one that that was particularly hot, particularly fluid and particularly capable of assimilating its wall rocks. The magma also contains a full complement of Ni and PGE, metals that are removed during the fractional crystallization that has affected the more evolved komatiitic magmas such as those that erupted in the other areas, including the Abitibi belt in Canada.

Box 3.3: Photos of the Norilsk-Talnakh Region

The town of Norilsk is situated at 56° N, in the far north of Siberia, very close to the Arctic circle. The deposits are fabulously rich, some comprising lenses of massive sulfide that are tens of metres thick and contain high concentrations

(continued)

of Ni, Cu and platinum-group metals. In terms of the total value of contained metals, the deposits are among the richest in the world.

The deposits at Norilsk were first found in the 1950s and were mined by prisoners of the Soviet Gulag. A series of smelters were built to refine the ores and for several decades their S-laden fumes devastated the surrounding countryside and damaged the health of everyone in the region. The town of Norilsk has been listed as one of the ten most polluted sites in the world, a description that was certainly merited in the past. It remains to be seen whether recent moves to clean up the mining and smelting operation will have any significant effect. The photos above show some scenes from the town and surrounding countryside – the damage to the buildings (bottom right) is due to only partly to the smelter fumes, being compounded by shoddy construction and the effects of the extremely harsh climate of northern Siberia.

3.4.4 Norilsk-Talnakh Nickel Sulfide Deposits

These remarkably large and rich deposits are located in northern Russia (Fig. 3.11a) in a tectonic setting that is very different from Kambalda. The deposits are hosted by small, shallow-level intrusions that form part of the enormous Siberian magmatic

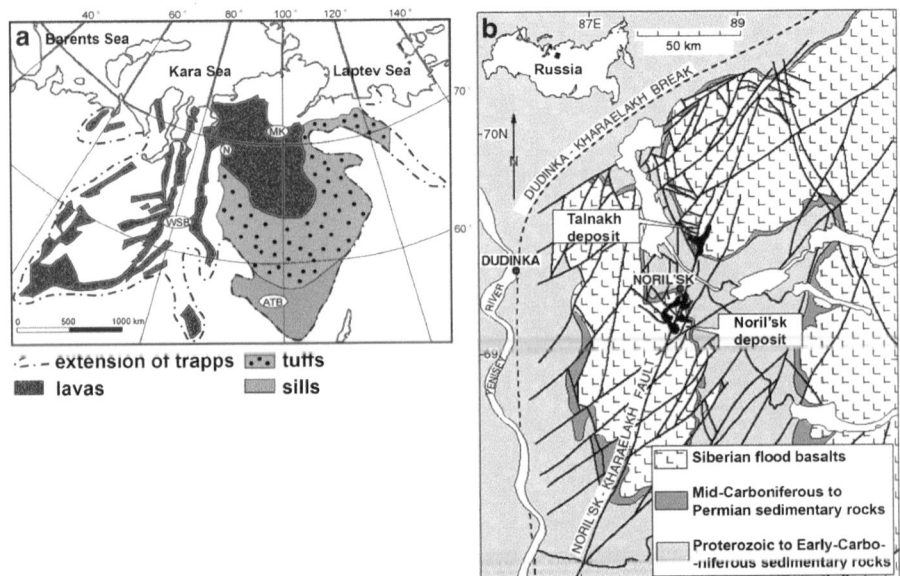

Fig. 3.11 Siberian flood basalts and the Norilsk-Talnakh Ni sulfide deposits

province. It is not entirely coincidental that one of the largest continental flood basalts hosts one to the largest ore deposits; but, as we shall see, this is not the whole story.

Figure 3.11b, a geological map of the region, shows the vast extent of the flood volcanic province – it covers an area similar to that of Western Europe. The deposits are located in the northern part of the province, at a place where later deformation has brought to the surface the base of the lava pile and the sedimentary rocks onto which they erupted. (Without this deformation the deposits would have remained several kilometres below the surface, hidden from prospectors and probably unmineable). The sedimentary sequence is invaded by a vast complex series of sills, as shown in Fig. 3.12, and these sills host the ore deposits. More detailed descriptions of the geology of the Norilsk regions and its ore deposits are provided by Naldrett (2004) and Czamanske et al. (1995).

A schematic section through an ore-bearing sill (Fig. 3.13) illustrates its complex geometry and lithology. The ore deposits are found in thicker-than-normal parts of the sill, and these segments are crudely differentiated, from olivine-enriched "picritic" lower portions to leucogabbroic upper portions. The ore occurs as remarkable, metre-thick layers of massive sulfide at the base of the intrusion, as disseminated sulfide in the interior, and as veins and lenses throughout the intrusion and penetrating into the wall rocks.

The mineralogy of the ores is similar to that at Kambalda, but they have higher Cu contents. This is related to the compositions of the magmas from which they

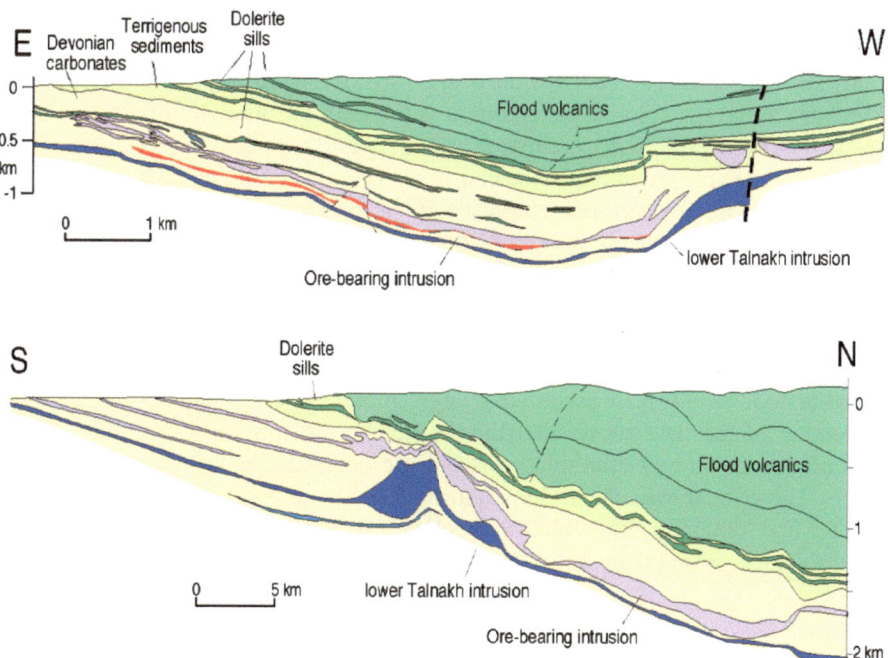

Fig. 3.12 Cross section through volcanic pile and underlying sill complex

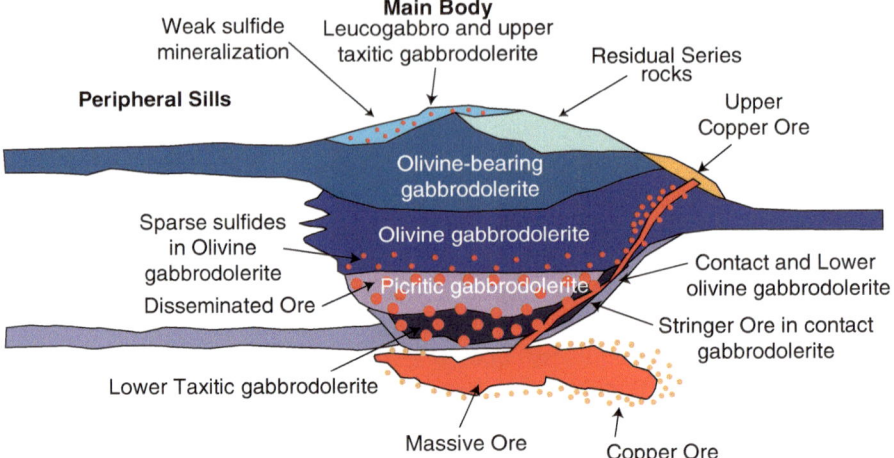

Fig. 3.13 Cross section through an ore-bearing sill

form; the Ni-rich, Cu-poor ultramafic magmas from Kambalda produced ores with Ni/Cu ratios of about 10 whereas the basaltic magmas at Norilsk, which have lower Ni and higher Cu, produced ores with ratios closer to 2.

In a broad sense the origin of the ores is like that at Kambalda. An immiscible sulfide liquid segregated from the silicate liquid and the dense droplets settled to the base of the intrusions. But what caused the sulfide to segregate from magma of basaltic composition? In addition, the Siberian large igneous province is only one of many such provinces, and despite the best efforts of mineral exploration companies who have actively explored the others, it is the only one known to contain a large magmatic ore deposit. Why is this?

Part of the explanation lies in the enormous volume of erupted lava, and the high magma fluxes involved in their emplacement. Recent age dating has shown that the vast majority of the lava pile was emplaced in a geologically short time period, most probably less than one million years, with the implication that an enormous volume of hot mantle melted rapidly, and that large amounts of hot magma flowed rapidly through the crust to the surface. Such circumstances favour crustal interaction, and indeed abundant geochemical data provide convincing evidence that many of the Siberian flood basalts have assimilated large amounts of continental crust. Yet close inspection of the data shows that the key to the ore forming process is not the assimilation of normal granitoid crust but a process that took place at shallower levels. The critical evidence is shown in Figs. 3.12 and 3.15.

The first diagram shows the sedimentary sequence that underlies the flood basalts and is invaded by the ore-bearing intrusions. The uppermost formation consists of Permian terrigenous sediments, the lower formations of Silurian to Devonian carbonates, marls, and evaporites (Fig. 3.14). It is commonly believed (e.g. Arndt et al. 2003; Li et al. 2009; Naldrett 1992) that these rocks played a

Fig. 3.14 Photos of Siberian flood basalts, ores and evaporite

crucial role in ore formation. The evaporites are made up of anhydrite ($CaSO_4$) a potential source of sulfur.

The second diagram (Fig. 3.15) compares the sulfur isotopic compositions of the Norilsk-Talnakh ores with those of uncontaminated mantle-derived magmas and with the likely composition of the sedimentary wall rocks. In the ores, the sulfur is isotopically heavy, with a composition approaching that of the evaporates, and very different

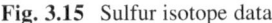

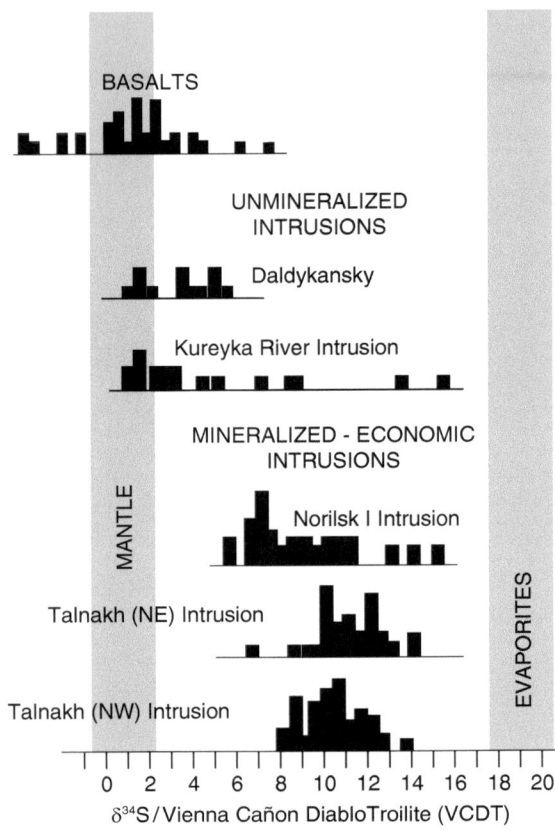

Fig. 3.15 Sulfur isotope data

from that of normal mantle magmas. In the opinion of many geologists, this sulfur was assimilated into the magma where it triggered the segregation of the sulfide ores.

The assimilation of sedimentary countries is believed to have taken place as magma flowed through conduits in the sedimentary pile, as illustrated in Fig. 3.16. To complete the picture it must be noted that the sedimentary S is assimilated in the form of sulfate. To convert it to sulfide requires the addition of a reductant, which, in the case of Norilsk was either coal of the upper sedimentary unit or organic matter in the carbonates.

3.4.5   Other Ni Sulfide Deposits

Of equivalent size to the Norilsk-Talnakh deposits are those of Sudbury in Canada. For many decades following its discovery at the end of the nineteenth century (see the account in Naldrett 2004), Sudbury was the only known major Ni deposit and it served as a model for the exploration of other deposits. This turned out to be a red herring because this deposit is truly unique, being the only known ore deposit

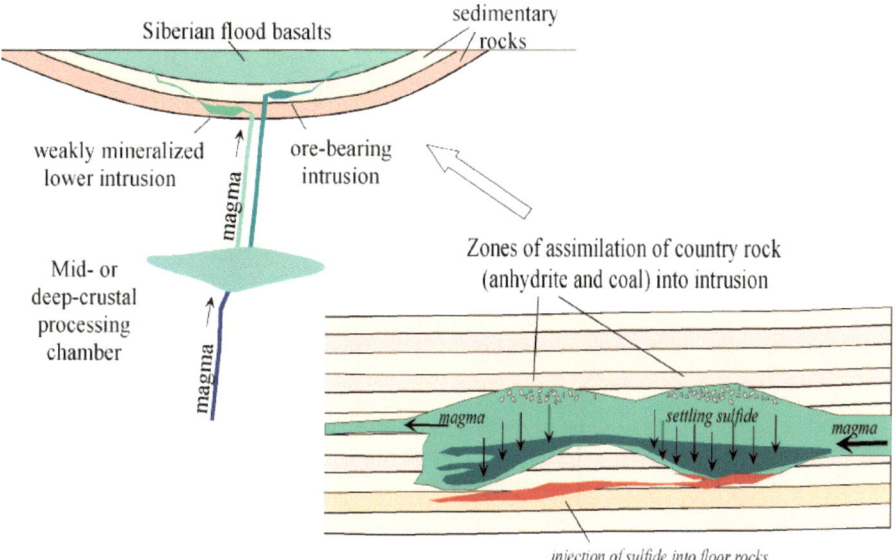

Fig. 3.16 The ore-forming process at Norilsk-Talnakh

associated with a meteorite impact. Most Sudbury ores occur as massive layers and pods in depressions in the lower contact of the Sudbury Irruptive Complex, a differentiated intrusion interpreted as the sheet of molten crustal rock that formed by total melting of the rocks at the site of impact (Fig. 3.17). Other deposits occur in "offshoots" the name given to vein-like intrusions extending outwards from the margin of the irruptive.

The impact that generated the Sudbury irruptive and its ore deposits was located at the contact between two crustal provinces, one an Archean granite-greenstone terrain, the other consisting mainly of sedimentary rocks. The melt sheet incorporated material from both, including a small fraction of sulfide from mafic intrusions that are inferred to have been present at the site of impact. The molten mafic material, being dense, accumulated in the lower part of the melt sheet, beneath an upper layer of molten felsic rock. The entire melt sheet was extremely hot (it was probably several hundred degrees above its liquidus) and had particularly low viscosity. Sulfide droplets that segregated from the molten rock were therefore able to settle efficiently to the base of the intrusion or were injected along fractures into the enclosing rocks. A more complete description of the Sudbury deposits is found in Naldrett (2004, Chap. 8).

Two other Ni-Cu sulfide deposits deserve mention, Voisey's Bay in Newfoundland and Jinchuan in China. The Voisey's Bay deposit is unusual in that it is hosted in mafic intrusions that form part of an anorogenic suite that includes troctolites and anorthosites (Li et al. 2000), but the ore formed in a manner broadly similar to that a Norilsk-Talnakh. As magma flowed up through a complex series of intrusions, it

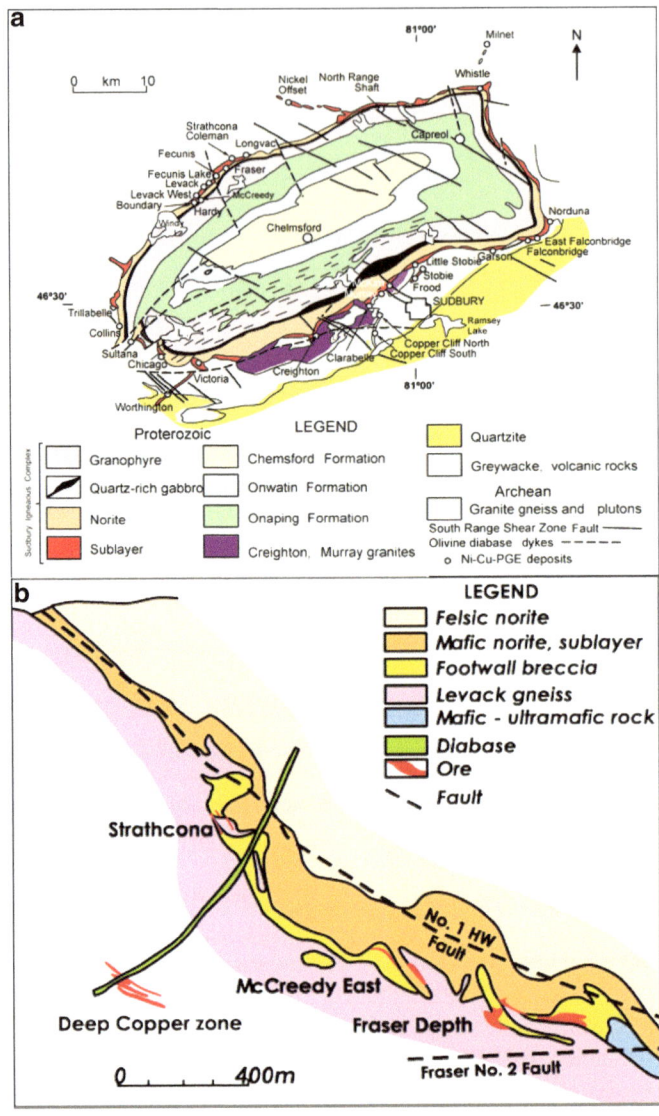

Fig. 3.17 The Sudbury irruptive complex (**a**) map, (**b**) section (from Eckstrand and Hulbert (2007) http://gsc.nrcan.gc.ca/mindep/synth_dep/ni_cu_pge/index_e.php

interacted with its wall rocks, perhaps picking up sulfur from sulfide-rich metasediments, a process that led to the deposition of Ni-bearing sulfides higher in the magma conduit. The intrusion that hosts the largest ore body is relatively small and it is clear that the sulfides could not have been derived from the small amount of magma represented by the intrusion. Evidently the sulfides were transported from elsewhere, probably from a deeper magma chamber, and were subsequently trapped in the narrow conduit where they are now found.

Jinchuan is the third largest Ni-Cu sulfide deposit (after Noril'sk and Sudbury). Unlike the others that we have described in this chapter, which all formed at or near the surface, the Jinchuan deposit is located at mid-crustal level in a series of strongly metamorphosed and highly deformed gneisses and marbles. The host intrusion is once again small (only about 6 km long and a few hundred metres wide) and it is composed almost entirely of olivine-rich ultramafic rocks. A conspicuous feature of the geological setting is the virtual absence of S-bearing country rocks, which seems to rule out assimilation of external sulfur as the ore-forming process. Tang (1993) and Lehmann et al. (2007) have suggested that the first stage of ore formation involved the contamination of komatiitic magma in a deeper staging chamber. Then a mush composed of olivine crystals and sulfide droplets in a silicate liquid was injected into the present Jinchuan intrusion. Sulfide segregation or transport may have been aided by interaction of the magma with wall-rock marbles.

3.5   Other Magmatic Deposits

Table 3.3 lists several other types of deposits that are found in igneous rocks and are thought to form mainly by magmatic processes. In fact it is not always straightforward to decide whether a deposit should be classed as magmatic or hydrothermal. Porphyry deposits (described in the following chapter) are the most important source of copper, and these deposits are indeed located within or adjacent to granitic rocks. However, as will be shown in the following chapter, they form through the precipitation of ore minerals from aqueous fluids and thus fit our definition of a hydrothermal deposit. In the case of tin deposits in granites (Table 3.3) similar ambiguity exists because many of these ore bodies are located at the marginal zones of granitic plutons where late-magmatic fluids have interacted with country rocks to produce a type of alteration, called greissen, which produces rock enriched in tin as well as other metals such as Sb, Cu, Pb, Zn. This type of deposit also lies at the limit between magmatic and hydrothermal. The most important magmatic tin deposits are located in the Malaysia, Indonesia, China, Australia, and Brazil. The South American countries Peru and Bolivia also produce large amounts of tin from polymetallic (Ag-Pb-Zn-Sn) hydrothermal deposits. Historically important deposits in Cornwall provided the metals that helped fuel Britain's industrial revolution.

Box 3.4 The Tin Fiasco of the 1980s

For much of the last century Malaysia was the world's major tin producer. In the 1980s it formed a cartel with other tin-producing countries to try to protect tin prices. Substitutes had emerged for traditional tin applications, particularly the use of protective plastic coatings inside what were once called "tin cans". This, together with increased recycling, had stifled demand for the metal.

(continued)

In 1981, the Malaysian government helped set up the International Tin Council, which bought up surplus tin stocks to maintain steady prices and in so doing to "protect the national interest" of Malaysia. This operation did indeed lead to a rapid price rise, from less than 7,000 to 9,000 £/t in 8 months. The Council then went further, by buying tin for cash in an attempt to control the global tin market. The purchases were supported by loans from banks linked to the Malaysian government. Crisis loomed when the cost of holding the tin became insupportable, and this pressure, together with the actions of market speculators, precipitating a massive collapse of tin prices. The new low tin prices rendered unviable tin mining operations throughout the world. Thousands of mine workers in Malaysia and elsewhere lost their jobs and the last tin mines in Britain were closed. The "good intentions" of the Malaysian government had backfired cruelly, and after the mid-1980s, tin was no longer the country's major export.

http://stocktaleslot.blogspot.com/2008/07/1981-2-malaysian-tin-market-fiasco.html

This story, together with other tales of misdirected government intervention, illustrates the perils of trying to manipulate world markets. OPEC, the petroleum cartel, has operated more or less successfully for four decades, largely because of the capacity of Saudi Arabia to increase or decrease production when the situation demands, so as to maintain the oil price at a level that finds a balance between the need to assure strong returns for producing countries and avoid stifling the world economy. The same does not apply to a commodity with only a small market like tin.

Less ambiguity surrounds the classification of deposits of Fe-Ti-V oxides in gabbroic or anorthositic intrusions. The example of the V-rich magnetite ores in the upper zone of the Bushveld Complex, which clearly formed through the accumulation of magmatic minerals in the upper part of the Complex, has already been discussed. Other examples, like the Fe-Ti-V oxides in gabbros of the Panxi region in China, form from Fe-rich magmas that intruded as part of the Emeishan large igneous province.

An important class of ilmenite (Fe-Ti oxide) deposits occurs in anorthosite massifs in Canada and Norway. Anorthosite is a rock consisting almost entirely of calcic plagioclase; anorthosite massifs were emplaced in a restricted time interval in the mid Proterozoic. Together with heavy mineral deposits in beach sands (Chap. 5), the deposits in these massifs are the dominant source of the high-technology metal titanium. The origin of these deposits is poorly understood – it is not clear how such large amounts of ilmenite, which normally is a late-crystallizing mineral, could have accumulated – but this lack of understanding is perhaps not so

Table 3.3 Other types of magmatic ore deposits

Commodity	Rock type	Geological setting	Examples
Tin	Granite	As a magmatic mineral (cassiterite) within granitic plutons and along margins where late magmatic fluids have interacted with country rocks (greissen)	Tin granites of Malaysia, Australia, Brazil
Iron, titanium, vanadium	Gabbroic and anorthositic intrusions	Ti-V-bearing magnetite occurs as a cumulus phase in gabbroic intrusions; large ilmenite deposits occur in Proterozoic anorthosite massifs	Fe-Ti-V: Bushveld, South Africa; Panzhihua, China Ilmenite: Tellnes, Norway; Allard Lake, Canada
Uranium	Leucogranite	Disseminated uraninite in leucogranite dykes	Rössing, Namibia
Lithium, beryllium, tin, tantalum, niobium, etc.	Pegmatite	Magmatic minerals in aqueous fluids released at the end stage of crystallization of granitic magma	Wodgina and Greenbushes, Australia: Bernic Lake, Canada; Marropino, Mozambique
Cu Zr, Ti, U, magnetite, vermiculite	Carbonatite	A carbonatite pipe in a alkaline pyroxentitic intrusion	Palabora, South Africa
Rare earth elements, Nb, Ta	Carbonatite	The REE occur in bastnäsite, $(Ce,La, Nd,\dots).CO_3F$, in carbonatite intrusions	Mountain Pass, USA; Bayan Obo, Mongolia
Diamond	Kimberlite, lamproite	Diatremes in Precambrian cratons	Numerous deposits in Botswana, Russia, Congo, South Africa, Canada, Australia

surprising because we have no entirely convincing model to explain the anorthosite massifs themselves.

Yet another type of magmatic ore deposit is found in pegmatites and carbonatites, from which metals like Li Be, B, Sn, Nb, Ta and the rare earth elements are mined. In this type of geological setting, little question surrounds the geochemical grounding for the association between metal and host rock. The metals in question are highly incompatible (i.e. they cannot be accommodated in the crystal lattices of common silicate minerals) and for this reason they become concentrated in highly evolved aqueous silicate liquids (e.g. pegmatite) or in the products of low-degree melting of the mantle (e.g. carbonatite). Deposits in pegmatites and carbonatites are generally small, but with accelerating industrial demand for high-technology metals such as Li and the rare earth elements, which are used in batteries or find multiple applications in the electronics industry, they are increasing sought after by mineral exploration companies and governmental agencies (see Chap. 6).

A notable exception to the small size of such deposits is the enormous Palabora deposit in South Africa, which is the country's largest copper deposit and a source of numerous other commodities, including Zr, Ti. U, apatite (fertilizer), vermiculite (a clay mineral used as an insulator or a growing medium in agriculture -those fluffy shiny grains you buy from your local garden store) and magnetite. The Palabora deposit is hosted by a carbonatite pipe within an alkaline pyroxenitic intrusion and the open-cast mine that was excavated in the carbonatite is said to be the largest man-man hole in Africa, if not in the world (see Fig. 3.18).

3.5.1 Diamond

Diamond in kimberlite is perhaps the best-known type of magmatic deposit. Kimberlite is a special type of ultramafic magma, one that is charged with volatile components such as water and CO_2. The magma is rich in potassium and incompatible trace elements and is probably produced by low-degree partial melting of a volatile-rich, geochemically enriched source deep in the mantle. Controversy exists as to whether this source is located in the basal continental lithosphere or at greater depths in the asthenosphere (Fig. 3.19a). Lamproite, a rock type similar to kimberlite, can also contain commercial diamond deposits. Kimberlites and lamproites are emplaced during highly explosive volcanic eruptions and form small, circular, funnel-shaped craters called maars (Fig. 3.19b). Most kimberlites are restricted to continents, and, according to Clifford's Rule (Clifford 1966), diamond deposits occur preferentially near or at the margins of stable Archean cratons. There are, however, some important exceptions, most notably the giant Argyle mine in Western Australia, which is located in a Proterozoic setting. This deposit is the greatest producer of diamonds, in terms of quantity but not value because the quality of the diamonds generally is poor. Unlike most other deposits, the Argyle mine is hosted by lamproite, not kimberlite.

Several decades ago almost all diamond mines were located in southern Africa but many large and important deposits have recently been found in Russia, Australia, and Canada. The small African country Botswana is now the world's second largest diamond producer (after Russia, measured by the values of the gems), followed by Canada. This is a remarkable turnaround because diamond deposits were quite unknown in Canada in 1990. Since then the discovery and development of two major deposits, Ekati and Diavik, both of which produce a large proportion of high-quality gems, has led to the country becoming a major force on the global diamond market.

Fig. 3.18   Two views of the Palabora open pit – a major Cu-rich multi-element deposit hosted in an alkali intrusion in South Africa

Strictly speaking diamonds in kimberlites are not truly magmatic. They are thought to be xenocrysts that were plucked from the sub-continental lithospheric mantle as the kimberlite magma ascended from its deep source to the surface. Diamond is the stable form of carbon at the pressures and temperatures that reign in the lower part the lithosphere. Given that carbon is relatively abundant in mantle rocks, it is probable that this part of the mantle is a vast reservoir of the gemstone. Kimberlite magma is merely a vehicle that transports the diamonds rapidly to the

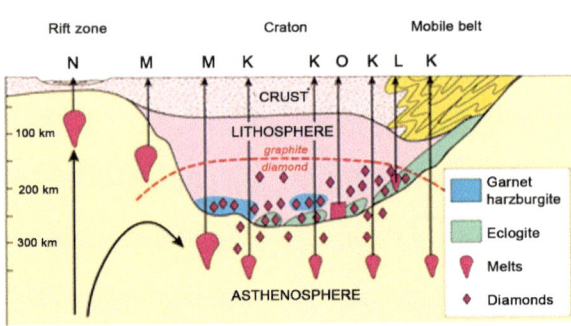

Fig. 3.19 (**a**) Lithosphere with diamonds; (**b**) Maar

surface under conditions that prevent them from reverting to graphite, their unattractive low-pressure polymorph.

References

Arndt NT, Czamanske GK, Walker RJ, Chauvel C, Fedorenko VA (2003) Geochemistry and origin of the intrusive hosts of the Noril'sk-Talnakh Cu-Ni-PGE deposits. Econ Geol 98:495–515

Ballhaus CG, Stumpfl EF (1986) Sulfide and platinum mineralization in the Merensky Reef: evidence from hydrous silicates and fluid inclusions. Contrib Mineral Petrol 94:193–204

Boudreau AE (1995) Some geochemical considerations for platinum-group element exploration in layered intrusions. Exploration Mining Geol 4:215–225

Campbell IH, Naldrett AJ (1979) The influence of silicate:sulfide ratios on the geochemistry of magmatic sulfides. Econ Geol 74:1503–1505

Campbell IH, Naldrett AJ, Barnes SJ (1983) A model for the origin of the platinum-rich sulfide horizons in the Bushveld and Stillwater complexes. J Petrol 24:133–165

Cawthorn RG (ed) (1996) Layered intrusions. Elsevier, Amsterdam

Cawthorn RG, Barnes SJ, Ballhouse C, Malitch KN (2005) Platinum-group element, chromium, and vanadium deposits in mafic and ultramafic rocks. Econ Geol 100th Anniversary Volume:215–249

Clifford TN (1966) Tectono-metallogenic units and metallogenic provinces of Africa. Earth Planet Sci Lett 1:421–434

Czamanske GK, Zen'ko TE, Fedorenko VA, Calk LC, Budahn JR, Bullock JH Jr, Fries TL, King BS, Siems DF (1995) Petrography and geochemical characterization of ore-bearing intrusions of the Noril'sk type, Siberia; with discussion of their origin. Resour Geol Special Issue 18:1–48

Eckstrand, O.R., and Hulbert, L.J., 2007, Magmatic nickel-copper-platinum group element deposits, in Goodfellow, W.D., ed., Mineral Deposits of Canada: Geological Association of Canada, Mineral Deposits Division, Special Publication No. 5, p. 205–222

Frost KM, Groves DI. 1989. Magmatic contacts between immiscible sulfide and komatiite melts; implications for genesis of Kambalda sulfide ores. Econ Geol 84: 1697–704

Groves DI, Korkiakkoski EA, McNaughton NJ, Lesher CM, Cowden A. 1986. Thermal erosion by komatiites at Kambalda, Western Australia and the genesis of nickel ores. Nature 319: 136–8

Irvine TN (1975) Crystallization sequences in the Muskox intrusion and other layered intrusions: II. Origin of chromite layers and other similar deposits of other magmatic ores. Geochim Cosmochim Acta 39:991–1020

Irvine TN (1977) Origin of chromite layers in the Muskox intrusion and other stratiform intrusions: a new interpretation. Geology 5:273–277

Lehmann J, Arndt NT, Windley B, Zhou MF, Wang C, Harris C (2007) Geology, geochemistry and origin of the Jinchuan Ni-Cu-PGE sulfide deposit. Econ Geol 102: 75–94

Lesher CM (1989) Komatiite-associated nickel sulfide deposits. In: Whitney JA, Naldrett AJ (eds) Ore deposition associated with magmas. Society of Economic Geologists, Dordrecht, pp 45–102

Lesher CM, Keays RR (2002) Komatiite-associated Ni-Cu-PGE deposits. Canad Inst Mining Metall Petrol 54:579–617 (Special volume)

Li C, Lightfoot PC, Amelin Y, Naldrett AJ (2000) Contrasting petrological and geochemical relationships in the Voisey's Bay and Mushuau intrusions, Labrador: implications for ore genesis. Econ Geol 95:771–800

Li C, Ripley EM, Naldrett AJ (2009) A new genetic model for the giant Ni-Cu-PGE sulfide deposits associated with the Siberian flood basalts. Econ Geol 104:291–301

Marston RJ, Groves DI, Hudson DR, Ross JR (1981) Nickel sulfide deposits in Western Australia: a review. Econ Geol 76:1330–1363

Mathez EA (1995) Magmatic metasomatism and formation of the Merensky reef, Bushveld complex. Contrib Mineral Petrol 119:277–286

Mavrogenes JA, O'Neill C (1999) The relative effects of pressure, temperature and oxygen fugacity on the solubility of sulfide in mafic magmas. Geochem Cosmochim Acta 63:1173–1180

Naldrett AJ (1989) Stratiform PGE deposits in layered intrusions. In: Whitney JA, Naldrett AJ (eds) Ore deposition associated with magmas. Society of Economic Geologists, Dordrecht, pp 135–166

Naldrett AJ (1992) A model for the Ni-Cu-PGE ores of the Noril'sk region and its application to other areas of flood basalt. Econ Geol 87:1945–1962

Naldrett AJ (2004) Magmatic sulfide deposits: geology, geochemistry and exploration. Springer, Heidelberg/Berlin

Tang ZL (1993) Genetic model of the Jinchuan Nickel-copper deposit. In: Kirkham RV, Sinclair WD, Thorpe RI, Duke JM (eds), Mineral deposit modelling. Geological Association of Canada Special Paper 40, Ottawa, pp 398–401

Chapter 4
Hydrothermal Deposits

4.1 Introduction

This important class of ore deposits is the source of most of the world's metals. Hydrothermal deposits provide almost 100% of our Pb, Zn, Mo, and Ag, 60–90% of our Cu, Au and U, as well as gemstones and industrial materials such as clay minerals and quartz. Hydrothermal deposits are diverse, being present in a wide range of geological settings and tectonic environments: some are closely associated with granitic intrusions, others form on the ocean floor and still others are in sedimentary basins. What all the deposits have in common is their origin via the precipitation of metals or ore minerals from hot aqueous fluids.

4.2 Key Factors in the Formation of a Hydrothermal Ore Deposit

To form a hydrothermal deposit requires: (1) a source of fluid, (2) a mechanism by which the metals or minerals are dissolved in the fluid, (3) a trigger of circulation of the fluid, (4) a mechanism that precipitates the metals or minerals (Fig. 4.1). We will now discuss each of these factors in turn before illustrating, through the description of five selected types of deposit, how they are related to ore formation.

4.2.1 Source of Metals

Some types of hydrothermal deposits are directly linked to plutonic rocks, usually granitoids, and in such cases it is evident that the ore metals are derived from the magmas themselves. The best examples are the so-called "porphyry-copper

N. Arndt and C. Ganino, *Metals and Society: an Introduction to Economic Geology*, DOI 10.1007/978-3-642-22996-1_4, © Springer-Verlag Berlin Heidelberg 2012

Fig. 4.1 Key factors that control the formation of hydrothermal ore deposits

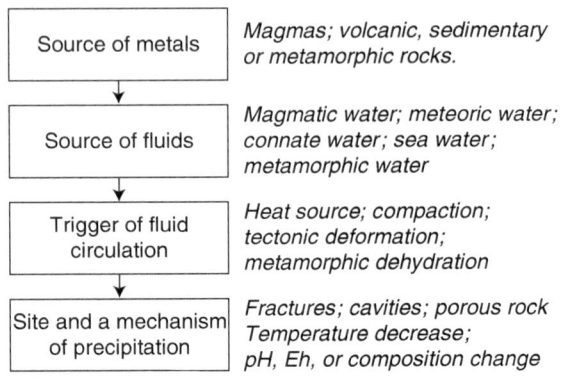

Key factors in the formation of a hydrothermal ore deposit

Source of metals	*Magmas; volcanic, sedimentary or metamorphic rocks.*
Source of fluids	*Magmatic water; meteoric water; connate water; sea water; metamorphic water*
Trigger of fluid circulation	*Heat source; compaction; tectonic deformation; metamorphic dehydration*
Site and a mechanism of precipitation	*Fractures; cavities; porous rock Temperature decrease; pH, Eh, or composition change*

deposits", a class that yields about 50% of the world's copper. For most other types of hydrothermal deposit, however, a link to magmas cannot be demonstrated and in these deposits the ore metals come from diverse rock types. In most cases they are leached from these rocks by the circulating hydrothermal fluids. Perhaps the best-known example are the volcanogenic massive sulfide or VMS deposits which form, and indeed are continuing to form at the present time, on the ocean floor. Seawater circulates through the oceanic crust, leaches out metals present at trace levels in the volcanic and sedimentary rocks of the crust, and reprecipitates them at the ocean floor to form the ore deposit. In other types of hydrothermal deposit, sedimentary or metamorphic rocks provide the ore metals.

The types of metals in the ore deposit are directly related to their source. The granitic source of porphyry copper deposits produces deposits that are rich not only in copper but also molybdenum, tungsten and in lesser quantities gold and silver. VMS deposits on basaltic crust are also rich in copper, in this case associated with zinc; when the substrate consists of felsic volcanic or sedimentary rocks, lead is present in addition to the two other metals. In deposits in a wholly sedimentary setting, copper is less present and the ores are dominated by lead and zinc. Finally hydrothermal gold and uranium deposits are hosted in a wide variety of crustal rocks and their origin depends crucially on the type of fluid, the manner in which the fluid circulates and the process that causes the metals to become concentrated.

4.2.2 Source and Nature of Fluids

The hydrothermal fluids that produce ore deposits are brines or more dilute aqueous fluids of diverse origins. Some also contain large CO_2 concentrations.

Using a variety of geological and geochemical techniques, particularly the analysis of fluid inclusions in ore and gangue minerals, it has been possible to identify the following types of fluid:

- Magmatic fluids released at various stages during the cooling and crystallization of granitic magmas
- Meteoric (rain) water
- Seawater
- Connate water, the interstitial water in pore spaces in sedimentary basins
- Metamorphic fluids, which are released by dehydration reactions in deeper crustal sections.

The temperatures of hydrothermal fluids, as determined from studies of fluid inclusions in ore and gangue minerals, range from more than 600°C in magmatic fluids to as low as 50–70°C for the fluids that deposit Pb-Zn sulfides or uranium minerals in sedimentary piles. Some examples are listed in Table 4.1.

The solubility of metals in pure water is very low, even at moderate to high temperature, and if the fluid is to be capable of transporting and re-depositing ore minerals, it must contain salt and other anions in solution. For many years it was very unclear how elements such as Pb or Au could be transported in hydrothermal fluids. Experiments carried out in the 1960s to 1980s yielded solubilities that were well below those required to form even modest-sized ore bodies. For example, the solubility of Zn in a slightly acid solution at 100°C is about 1×10^{-5} g L^{-1}. If such a fluid were to form an ore deposit, the amount of fluid that must pass through the deposit is about 100,000 km^3, a volume greater than that of the Caspian Sea! More recent work has shown that the solubility of ore metals is greatly increased in saline solutions by the formation metal complexes. The major natural ligands are hydroxide, Cl, S and in some cases HCO_3/CO_3 and F. Figure 4.2 shows how the concentration of Zn increases by more than 5 orders of magnitude as the Cl content increases from the low values present in rain or sea water to the higher values present in saline fluids. In this case the high solubility results from the formation of high-order chloride complexes; in other fluids complexes with various S species are important. The nature of metal complexes depends on the metal-ligand chemical affinities: so-called "hard metals" like Al, REE, Zr, U, preferentially form complexes with ligands like OH, F, and CO_3, whereas "soft metals" like Au, Pt strongly prefer HS or H_2S. Most base metals will be complexed largely with Cl. Discussions of these issues are found in Brimhall and Crerar (1987) and Chenovoy and Piboule (2007).

The compositions of hydrothermal fluids are listed in Table 4.1. The source of the Cl and S in these fluids was initially seawater, but to explain the high concentrations it is thought the Cl and F were introduced into the fluids indirectly by dissolution of evaporates and the sulfur by interaction with sedimentary sulfate or with reduced species such as pyrite and other diagenetic sulfides or with S-bearing organic materials or sour gas.

Table 4.1 Composition of selected hydrothermal fluids (Compilation from Chenevoy and Piboule 2007)

(a) Continental	T(°C)	pH	Na + K (ppm)	Ca (ppm)	Cl (ppm)	SO₄ (ppm)	SiO₂ (ppm)	CO₃ (ppm)	H₂S (ppm)	Cu (ppm)	Pb (ppm)	Zn (ppm)
Broadlands-Ohaaki, New Zealand (Simmons and Browne 2000)	260	6.3	541	7	25	19	170	1,144				
Waiotapu, New Zealand (Hedenquist and Henley 1985)	220	5.9	809	10	732	102	353	1,074	86			
	250	6										
Rotokawa, New Zealand (Krupp and Seward 1987)	339	7.1	563	2.5	871	35.8	436	496	27		50	100
	318	5.7	583	1.6	807	5.2	580	3,788	228		40	25
	329	6.7	355	1.1	515	11	579	4,575	106	25	175	125
Matsao, Taiwan (Ellis 1979)	245	2.4	6,390	1,470	13,400	350	369	2			1	13
Salton Sea, California (Ellis 1979)	340	5.5	77,800	40,000	184,000	10				8	102	540

(b) Oceanic	T(°C)	pH	Na + K (ppm)	Ca (ppm)	Cl (ppm)	SO₄ (ppm)	SiO₂ (ppm)	CO₃ (ppm)	H₂S (ppm)	Cu (ppm)	Pb (ppm)	Zn (ppm)
East-Pacific Ridge 21°N, NGS and HG (Von Damm 1990)	273	3.8	12,736	832	20,555		975		224	1.3	37.8	2,600
East-Pacific Ridge 13°N and 11°N (Von Damm 1990)	351	3.3	1,121	468	17,608		780		286	2,800	74.3	6,760
	354	3.1	13,745	2,148	25,276	37	966		279		5.6	325
Mid-Atlantic Ridge, TAG and MARK-1 (Von Damm 1990)	347	3.1	12,104	900	19,986		940		292		21.7	6,825
	321		14,095	1,040	23,394		1,100					
Juan de Fuca Ridge, Axial Volcano Inferno (Von Damm 1990)	350	3.9	12,650	396	19,844		910		201	1,088	10.3	3,250
	323	3.5	12,570	1,872	22,187		755		238	768	23.4	7,435
Juan de Fuca Ridge, Vent-1 (Von Damm 1990)	285	3.2	16,658	3,388	31,808		1,140		192	128		39,000
Gulf of California, Guaymas, 4 and 5 (Von Damm 1990)	315	5.9	12,719	1,360	21,264		690		163	70	47.6	1,235
	287	5.9	12,905	1,236	21,200		620		139	6	4.1	143

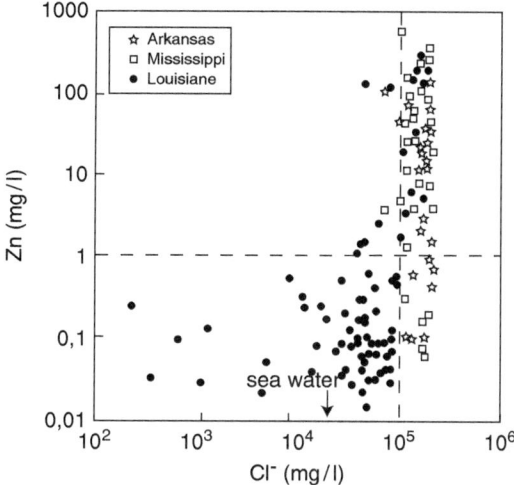

Fig. 4.2 Solubility of zinc as a function of chlorine of the hydrothermal fluid. The solubility increases dramatically when the Cl content is higher than 105 mg L^{-1} through the formation of chlorine complexes. If the Cl content decreases, for example when the fluid is diluted, Zn precipitates and can form deposits (Modified after Cathles and Adams 2005)

4.2.3 The Trigger of Fluid Circulation

In past decades a process called lateral secretion was discussed as a possible ore-forming process. The idea was that diffusion of metals and other elements along a thermal or chemical gradient could lead to the precipitation of metals in a restricted location and thus to the formation of an ore body. This idea has since fallen out of favour because it has been recognised that only under exceptional circumstances can elements diffuse up a chemical gradient, as is needed if elements present in low concentrations in a solution can precipitate as a high-concentration ore body. It is now recognised that most ore deposits form as a result of the circulation of hydrothermal fluids, and indeed that high fluid fluxes are required to form large ore bodies. An important question therefore is the nature of the process or motor that causes the fluid to circulate.

As for the other parameters discussed above there are several possibilities. For deposits related to magmatic activity release of fluids from magmas is the main driving force. The exsolution of fluid entails a large increase in volume which is capable of fracturing the rocks overlying the magma chamber and the low-density fluids thus released ascend through the fractures. Heating of groundwater surrounding the intrusion causes it to convect, enhancing the primary circulation.

Convection is the main cause of circulation in deposits that form at the ocean floor. Seawater penetrates into the crust where it acquires heat from still-hot lavas or from high-level intrusions. The less-dense warm fluids then ascend to the surface along fractures. In the case of deposits in sedimentary basins the driving mechanism is less obvious. Evacuation of pore fluids during compaction of sediments no doubt plays a role, but in many cases this is too slow a process to produce the sudden fluxes of warm fluids that are implicated in the formation of certain deposits. Tectonic loading associated with mountain building at the basin margins is implicated in some examples, but again this process is too slow to have the required effect. To form the Pb deposits in carbonate sequences more obscure processes such as the influx of warm seawater onto and thence into the basin has been suggested. As for the cause of the influx, changes in ocean circulation or melting of continental glaciers have been proposed.

4.2.4   A Site and a Mechanism of Precipitation

The crux of the ore-forming process is the precipitation of the ore metals or minerals. What is required is a mechanism that causes the metals to come efficiently out of solution and concentrate in a restricted volume of rock. The most common cause of precipitation is cooling of the solution, which decreases the solubility of the metals. Cooling takes place when hot magmatic fluids enter cool wall rocks, when fluids emerging from a seafloor spring mix with cold seawater or when warm basin fluids mix with cooler near-surface waters. Associated with many of these cooling events is dilution of the hydrothermal brines and since this decreases the concentration of the complex-forming anions, this also decreases the solubility of the metals, leading to their precipitation. Another process is reaction with wall rocks, which changes the fluid composition. Particularly important are redox reactions, which happen when oxidized basin waters come into contact with reduced materials such as hydrocarbons or organic-rich shales. This type of inter-action is crucial in the formation of most uranium deposits and many of the base-metal deposits in sedimentary basins.

There are two main types of deposition site; open fractures and zones of replacement. Many hydrothermal deposits form at shallow levels in the crust where fractures remain open and in such cases much of the mineralization consists of ore minerals that precipitated in such fractures. Ore bodies formed this way consist of a multitude of veins and patches of ore minerals dispersed through the host rock. Cavities and caves in limestone reefs and kast facies, and interstitial space in breccias, are important sites of deposition of Pb sulfides. And finally the ocean water in which the sulfides of VMS deposits accumulate could be considered a special case of open cavity precipitation.

Hot hydrothermal fluids are chemically aggressive and capable of reacting with a wide range of rock types. Alteration zones surround most hydrothermal systems and ore minerals occur in many of these zones. In some cases the minerals are disseminated or restricted to veins; in other cases wholesale replacement of the original rock is evident.

4.3 Examples of Hydrothermal Deposits and Ore-Forming Processes

We have selected five types of deposit to illustrate how different types of fluid in diverse geological settings can lead to the formation of an ore body. The list is by no means exhaustive – in a short text such as this it is impossible to describe the vast range deposits that form as the result of circulation of hydrothermal fluids. However our selection will, we hope, suffice to illustrate the essential features of this class of deposits. As with the description of magmatic deposits the emphasis is not on the characteristics of the deposits themselves but more on the processes that produced them.

4.3.1 Volcanogenic Massive Sulfide (VMS) Deposits

We start with this type of deposit because they are among the best understood of all ore deposits. There are various reasons for this: the ore bodies are relatively simple, both in their structure and their composition and mineralogy, and they have also been studied intensively over the last decades. But more to the point is the fact that they are one of very few deposits whose formation, by way of precipitation of sulfides at or just below the ocean floor, we can observe directly (other examples of active ore formation include the accumulation of heavy minerals in placer deposits and the accumulation of sulfidic sediments in sedimentary basins. These are described in Chap. 5).

The discovery in 1977 by scientists in the Alvin submersible of active hydrothermal vents – black smokers – on the ocean floor is one of the most important advances in earth (and biological) sciences of the past decades. The discovery has had profound implications for the origin and evolution of the oceanic crust and for the biological sciences, and it also opened a window through which we can study, in real time, the processes that generate an ore body. At each hydrothermal vent, sulfides rich in Zn, Cu and Pb precipitate in the chimneys that build up around each upwelling jet of hydrothermal fluid. The same sulfides separate out from the hydrothermal plume and settle onto the ocean floor. Most accumulations of sulfide minerals on the modern sea floor are relatively small but the long-lived system that built the TAG mound on the Mid-Atlantic ridge is estimated to contain about three million tons of sulfide grading 2% Cu with smaller concentrations of Zn and Au. If such a deposit were present on land (and not in a region hostile to mining), it certainly would be exploited.

VMS deposits were among the first ever to be mined: ores on Cyprus and in Spain, for example, were exploited over 2000 years ago and provided much of the copper used in the bronze weapons of Roman centurions. In the early part of the last century, when the opinions of American geologists like Lindgren held sway, these deposits were interpreted as epithermal replacement bodies produced by the

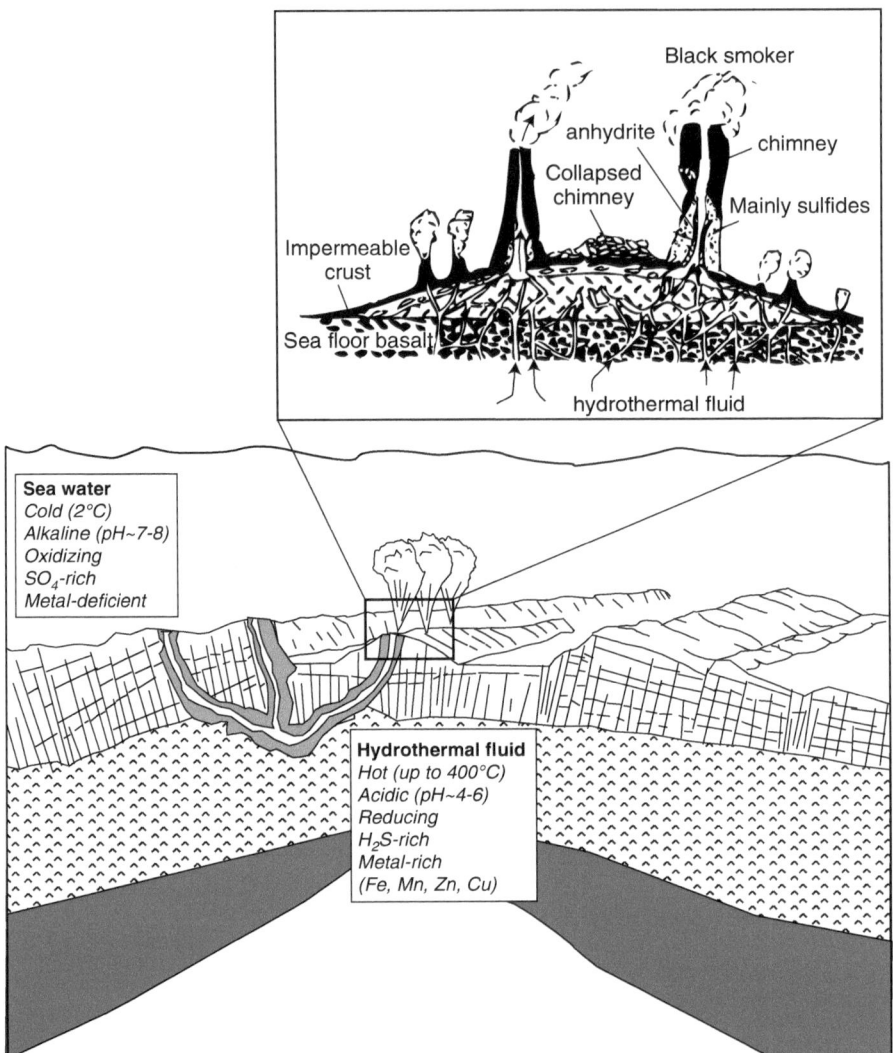

Fig. 4.3 Characteristics and general pattern of circulation of fluids at mid-ocean ridges. These fluids are responsible for the construction of black smokers and lead to the accumulation of sulfides on the seafloor (Modified from Robb 2007)

precipitation of sulfides from granite-sourced fluids. However, in the 1950s and 1960s geologists in Norway, Canada, and Australia developed the hypothesis that these deposits in fact had formed on the ocean floor, an idea vindicated by the subsequent discovery of the black smokers (Fig. 4.3). Strengthened by important contributions from Japanese geologists who undertook detailed studies of the Besshi deposits around the same time, a volcanic exhalative model for the formation of this class of deposits is now widely accepted.

Description: The distinctive feature of a VMS deposit is its association with volcanic rocks. Depending on the particularly setting, these can be mafic (basaltic) or felsic. Invariably they were deposited under water, either in a mid-ocean ridge setting or more commonly in island arcs or during arc-continent collision. Sedimentary rocks form an important part of the host sequence in certain classes of VMS deposits.

Most VMS deposits are relatively small, usually containing only a few million tons of ore. Exceptions are the large Kidd Creek deposit in Ontario, Canada (160 mt) and even larger deposits in Spain and in the Russian Urals (Box 4.2). The grades of the ores are high, however, which makes this type of deposit an attractive exploration target, particularly for small or "junior" mining companies. Typical grades are 1–5% of Cu, Zn, and/or Pb with minor quantities of Au and Ag.

The mineralogy is relatively simple. As with almost all sulfide deposits, with the exception of those in purely sedimentary settings, iron sulfides predominate. In VMS deposits, pyrite or pyrrhotite make up about 90% of the sulfide assemblage, which may also include chalcopyrite, sphalerite and galena, and in some cases and in minor amounts, bornite, arsenopyrite, magnetite, and tetrahedrite.

Most deposits have the very distinctive structure illustrated in Fig. 4.4. A tabular or mound-shaped body of stratiform, banded, massive sulfide overlies a crudely pipe-shaped discordant "stockwork", a zone of mineralized veins that cuts

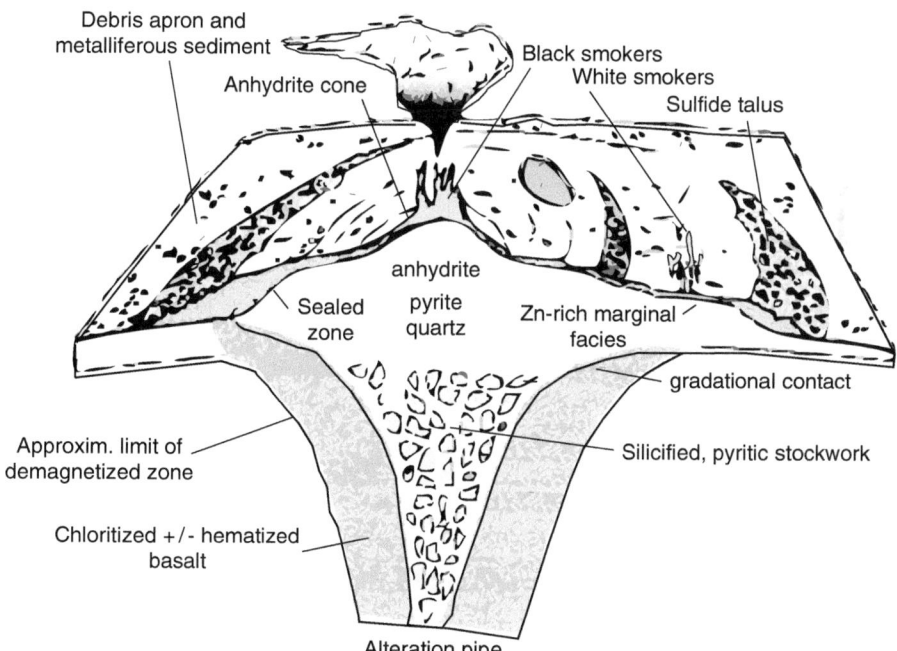

Fig. 4.4 Diagram of a typical VMS deposit, from the example of "TAG sulfide mound" on the media-Atlantic Ridge (Modified from Hannington et al. 1998)

Fig. 4.5 Photos of section cut through a chimney of a black smoker VMS-type deposit (Photo N. Arndt); (b) and (c) photos of ore from deposits in the Yaman Kasy massive sulfide deposit in the Urals. (b) is a fossilized tube worm dwelling tube and (c) contains fossilized monoplacopherans and brachiopods (photos from Phil Crabbe)

vertically through highly altered host rocks. The proportions of metals vary within the deposit: the upper massive sulfide is rich in Zn and Pb (in those deposits that contain this metal) whereas the stockwork is enriched in Cu and Au. In many deposits later deformation and metamorphic recrystallization has destroyed the original ore textures, but well-preserved examples preserve bedding and other sedimentary structures. In the remarkable deposits from the Urals and Ireland, the chimneys of black smokers are beautifully preserved, to the extent that even the dwelling tubes of tubeworms and other fossils, now replaced by sulfide, can be recognised (Fig. 4.5 and Box 4.2).

The volcanic rocks that host VMS deposits erupt in a wide variety of tectonic environments. Although the modern black smokers that provided the clues to their origin are best known along mid-ocean ridges, many more recent discoveries, and the locations of the ore deposits themselves, are in convergent margin settings. From the recent classification based on the rock types associated with the deposits, given in Table 4.2, we can see that the setting varies from intra-oceanic arc and backarcs through continental margins to mature epicontinental backarcs. The age spans most of geological history: the Big Stubby deposit in the Pilbara of Australia has an age of 3.5 Ga and is one of the oldest known ore deposits; large and important

Box 4.1 Examples of Modern and Ancient VMS Deposits

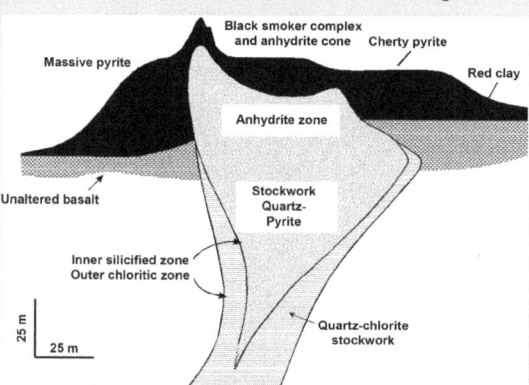

TAG sulfide deposit. This deposit is currently forming on the rift valley of the Mid-Atlantic ridge at 26°N, the site of an field of active black and white smokers (hydrothermal springs). The deposit is located on the seafloor above pillow basalts of the oceanic crust. It has the form of a classic VMS deposit, comprising an upper lens of massive and semi-massive sulfide underlain by a vertical pipe-like stockwork. Anhydrite, chert and red clay co-precipitated with the sulfides. The deposit contains 3.9 mt of ore, 2.7 mt of massive and semi-massive sulfide (~2% Cu) and 1.2 mt of mineralized breccias (~1% Cu) in the stockwork

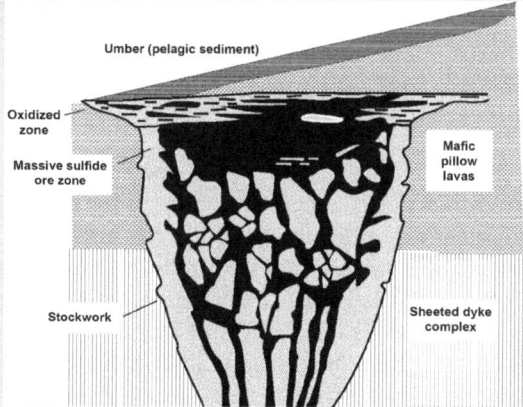

Cyprus VMS deposits. The Troodos ophiolite contains clusters of VMS deposits in pillow basalts that probably erupted in a Cretaceous back-arc basin. All deposits are Cu-rich (1–4%) with similar Zn tenors, as is normal for deposits hosted by mafic volcanic rocks. They are made up of a massive tabular cap overlain by sandy-textured and brecciated ore, in which the pyrite-rich massive core is cemented by chalcopyrite and sphalerite, and an underlying stockwork formed of a mixture of quartz and pyrite, with minor

(continued)

amounts of base metal mineralisation. Layers of *umber*, Fe-, Mn-, and trace-metal enriched mudstones of volcanic exhalative origin, cap the sulfide lens.

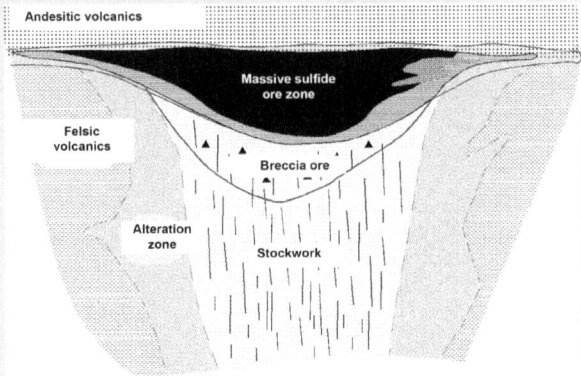

Archean VMS deposits. The Delbridge deposit is typical of VMS ore bodies in the 2.7 Ga Abitibi belt in Canada. The deposit formed at the contact between felsic pyroclastic rocks and andesitic volcanic lavas that were once part of an ancient island arc. The diagram to the left, redrawn from Boldy (1968), illustrates all the essential features of a VMS deposit and shows that Canadian geologists understood how these deposits formed well before the discovery of active black smokers in 1975. According to Boldy, the deposit is of "*volcanic exhalative origin*", an example of "*mineralization of a flank fissure which was the site of solfataric activity within which the various metals were rhythmically precipitated.*"

deposits also form in the late Archean and through the Proterozoic and Phanerozoic; and as mentioned above, deposits continue to form on the modern ocean floor.

The types of ore metals are directly related to the geological setting and host rocks. Ore bodies in mainly basaltic rocks are rich in Cu and contain only minor amounts of Zn and other metals, which leads to a parallel classification in which they are known as Cu-Zn deposits. Those in bimodal mafic-felsic settings are richer in Zn than Cu (Zn-Cu deposits); and those in sedimentary settings contain Pb in addition to the Cu and Zn (Zn-Pb-Cu deposits).

Origin: A genetic model for the formation of a VMS deposit, summarized from Franklin et al. (2005), is illustrated in Fig. 4.6. It has six main elements. (1) a heat source to drive the hydrothermal convective system and potentially to contribute some ore metals. In many deposits the source is a shallow-level intrusion of mafic to felsic magma; (2) a zone of high-temperature reaction in which metals and other components are leached from volcanic and/or sedimentary by circulating seawater; (3) synvolcanic faults or fissures which focus the discharge of hydrothermal fluids; (4) footwall, and less commonly, hanging-wall alteration zones produced by interaction

Box 4.2 VMS Deposits of the Urals

The Urals in Russia host six enormous VMS deposits, each containing more than 100 mt of ore, and many smaller deposits. These deposits formed in volcanic-dominated sequences that evolved during the Silurian and Devonian as oceanic island arcs collided with the Precambrian continent of central Russia. The deposits contain different metals that can be correlated with both the nature of the associated volcanic rocks and their tectonic setting. In certain deposits the dominant ore metals are Cu and Zn, and these are associated with tholeiitic mafic volcanics that erupted in an early arc setting. Other deposits are polymetallic and contain significant concentrations of Pb, Ag, and Au in addition to Cu and Zn. These are hosted by bimodal mafic-felsic calc-alkaline volcanics and sediments in forearcs or rifted arcs that developed during arc-continental collision.

A remarkable feature of many deposits is their excellent preservation due to an absence of metamorphism and deformation following their deposition. This has meant that their textures, structures and compositions show minimal disturbance, which has provided a window to the ore-forming processes. Herrington et al. (2005) have described how clastic and hydrolytic processes that preceded diagenesis on the ancient sea floor modified the morphology and mineralogy of the deposits. The excellent preservation also allowed the preservation of fossilized tube worm stuctures and other examples of the fauna that constitute part of the unique ecological systems surrounding hydrothermal vents. The similarity between the Silurian fossils and modern vent fauna attests to the slow evolution of this part of the biosphere.

between ascending hydrothermal fluid and seawater; (5) the massive sulfide deposit itself, formed at or near the sea floor; and (6) bedded sediments formed by precipitation of sulfides and other components from the hydrothermal plume.

A VMS deposit forms in the following way. Magma intrudes at a shallow level in the oceanic crust. It heats seawater that is present in pores and fractures in the volcanic and sedimentary rocks and causes the water to circulate through the volcanic pile (Fig. 4.6). As it does so it draws down seawater into rocks flanking the intrusion, thus setting up a convective system. The cold seawater percolates down through the oceanic crust through open fissures and the slightly alkaline water precipitates its sulfates and carbonates as it descends. Its temperature progressively increases and as the fluid approaches the magma chamber at 2–3 km depth, it has been transformed to hot hydrothermal fluid whose temperature is 350–400°C and whose pH has decreased to 4–6. As it approaches the critical point its volume increases drastically, driving it back up towards the surface. The hot, acid, corrosive liquid leaches metals from the volcanic or sedimentary rocks and these metals are transported upwards, probably as metal halide complexes. The fluids ascend along fractures until they reach the seafloor. On expulsion they cool rapidly and mix with

Table 4.2 Different types of VMS deposits

Type	Lithological association	Tectonic setting	Metals	Examples
Bimodal-mafic	Dominantly mafic volcanic but with up to 25%felsic volcanic stata	Volcanic arcs (rifted) above intra-oceanic subduction zones	Cu-Zn	Noranda, Abitibi belt, Canada – Archean; Flin-Flon, Canada – Proterozoic; mid and south Urals, Russia – Phanerozoic
Mafic	Ophiolite sections of basaltic lavas with minor boninite, cherts and mafic tuffs	Mature intra-oceanic back-arcs	Cu- (Zn)	Southern Urals; Newfoundland; Troodos in Cyprus
Pelitic-mafic	Basaltic lavas and sills and equal or greater amounts of pelitic sedimenary rocks	Mature, juvenile and accreted backarc	Cu-Zn-Pb	Outokumpo, Finland – Proterozoic; Windy Craggy, Canada – Paleozoic; Besshi, Japan – Mesozoic
Bimodal-felsic	Felsic volcanic and terrigenous sedimentary rocks in near-equal proportions	Continental margin arcs and related backarcs	Cu-Zn-Pb	Bergslagen, Sweden – Proterozoic; Tasman orogen, Australia – Paleozoic
Siliciclastic-felsic	Felsic volcanoclastic rocks and high-level intrusions; minor mafic lavas and chemical sediments	Mature epicontinental backarcs	Cu-Zn-Pb	Golden Grove, Australia – Archean; Iberian pyrite belt, Spain and Portugal; Bathurst, Canada

cold seawater, which drastically decreases the metal solubility, leading to the precipitation of metal sulfides, together with barite, anhydrite and silica. Some of the sulfides accrete around the hydrothermal vents to build chimneys that reach 10's of metres high before they crash down to form a layer of sulfide debris mixed with precipitated chemical sediment on the seafloor. This layer has low permeability and hinders the ascent of fluid to the surface; the trapped fluid accumulates beneath the seafloor where it mixes with seawater and precipitates more sulfide. In this manner the main tabular or lens-shaped body of massive sulfide is built up. Some of the fluid escapes to form a hydrothermal plume that ascends many hundreds of metres around the black smoker and precipitates sulfide particles that settle out to form bedded "exhalative" sediments around the site. The stockwork beneath the ore body forms as high-temperature hydrothermal fluid interacts with wallrocks and seawater in the conduits that transfer the fluid to the surface.

Fig. 4.6 Photographs of (**a**) the open pit of the Chiquicamata porphyry copper deposit in northern Chile, and (**b**) the open pit of the Exotica deposit near the main Chiquicamata deposit (Photos N. Arndt)

The order in which the sulfides precipitate depends on solubility and temperature. Copper and Au react out at high temperature in the plumbing or stockwork beneath the chimney and Fe precipitates as pyrite at the base of the pipe. The chimney itself, as well as the 'smoke' from a black smoker, is composed of Zn and Pb sulfides, as well as barite and anhydrite.

Analysis: VMS Deposits
Source of metals – volcanic and sedimentary rocks of the oceanic crust
  Source of S – seawater sulfate
  Source of fluid – seawater
  Cause of fluid circulation – convection, commonly related to high-level magma chambers
  Precipitation process – cooling, change in redox state and dilution as hydrothermal fluid reacts with seawater.

4.3.2 Porphyry Deposits

Introduction: Porphyry deposits are the world's most important source of Cu and Mo, and also produce significant amounts of Au, Ag, W, and Sn (Sinclair 2007; Sillitoe 2010). They account for about 50–60% of world Cu production and more than 95% of world Mo production. In contrast to VMS deposits, which normally are small (1–5 mt) but of high grade (3–10% ore metals), porphyry deposits are enormous but of low grade. The best-known deposits are in the cordillera of North and South America, the location of the Bingham ore body in the USA (2,733 million tons of ore grading 0.7% Cu and 0.05% Mo) and the Chuquicata ore body in Chile (10,837 mt of 0.56% Cu and 0.06% Mo) (Box 4.3). The latter deposit is the site of what is said to be the world's biggest open-pit mine and its neighbour, the El Teniente deposit, is exploited in the biggest underground mine. Sinclair's (2007) compilation lists 44 deposits with reserves greater than one billion tons of Cu, Mo, or Au ore.

Another large deposit is the Grasberg ore body in Irian Jaya, the Indonesian (western) portion of New Guinea, which contains about 2,100 mt of ore grading 1.2% Cu and 1.2 g/t of Au, making it, on one hand, the biggest gold mine and the third biggest copper mine in the world, and on the other, the site what has been described as the "world's worst eyesore". The conflict created by the environmental damage engendered by an enormous mining operation in a region of fragile, high-altitude rain forest and the immense economic benefit of the operation, which contributes 2% of the entire gross domestic product of a very poor country, starkly illustrates the dilemma associated with the exploitation of the Earth's natural resources.

Structure and Mineralization: Returning to geological issues, porphyry deposits derive their name from the phenocryst-bearing felsic to intermediate shallow-level intrusions with which they are associated. The form of porphyry deposits is highly varied and includes irregular, oval, solid, or "hollow" cylindrical and inverted cup shapes. As shown in Fig. 4.8, the ore bodies are superimposed on the upper parts of relatively small granitic plutons which represent offshoots from larger batholiths at greater depths. The shallow-level plutons are located in the lower portions of volcanoes and no doubt are parts of conduits that supplied magma to overlying volcanism. Both the ore bodies and the plutons are composite structures built up a many individual pulses of magma and hydrothermal fluid. The ores are not confined to the plutons but extend outwards into the surrounding rocks (Fig. 4.7). Closely associated with the mineralization is moderate to intense alteration which displays a zoning concentric about the pluton. This alteration also extends well outside the zone of mineralization and is used as a guide during the exploration of this class of deposits. The zoning in the alteration in the deposits from western USA, which are considered a classic type of porphyry deposit, is illustrated in Fig. 4.8.

The mineralization consists of small concentrations of sulfide minerals, disse-minated or dispersed in small veins and replacement patches in the highly altered

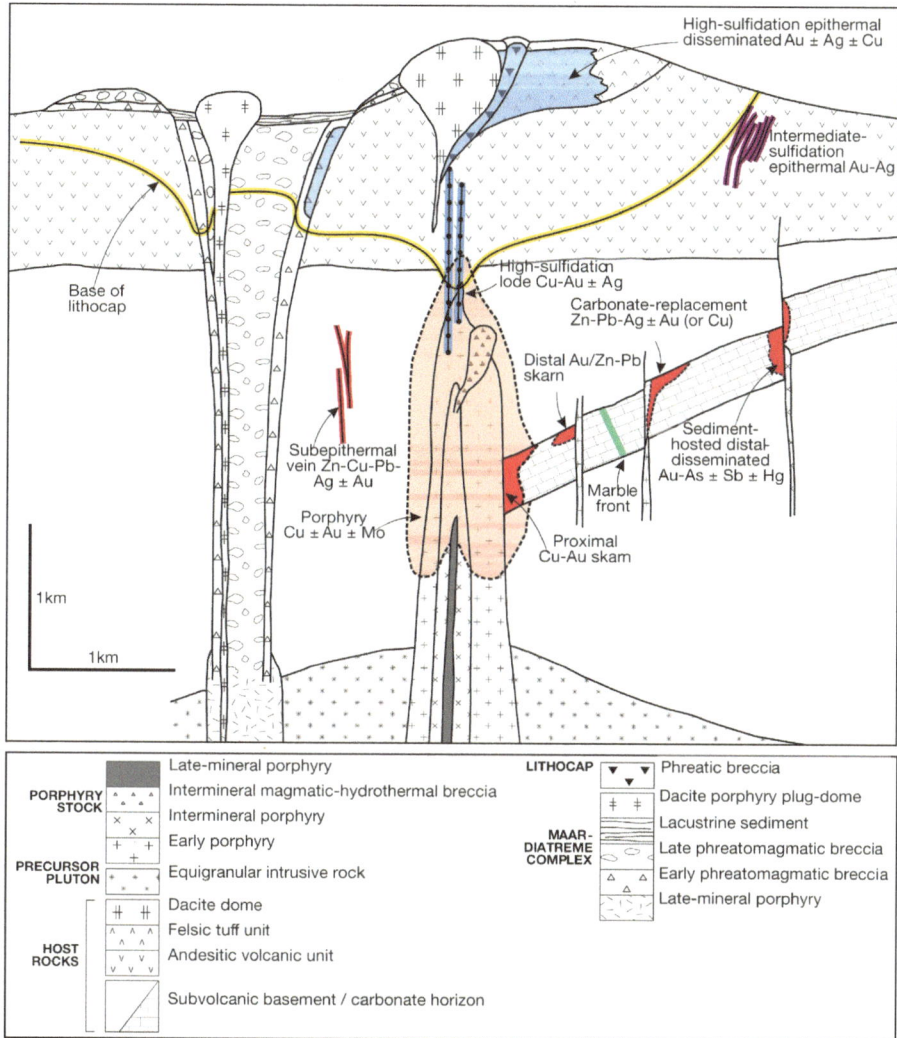

Fig. 4.7 Diagram of the upper part of a granitic pluton in a volcanic edifice, the location of many porphyry copper deposits (Modified from Sillitoe 2010)

upper portions of the intrusion and in surrounding rocks. Original sulfide minerals are pyrite, chalcopyrite, bornite, and molybdenite. Gold is often in native form and is found as tiny blobs along borders of sulfide crystals, or it occurs in sulfosalts like tetrahedrite. Most of the sulfides occur in veins or plastered on fractures and most are intergrown with quartz or sericite. In many cases, the deposits have a central low-grade zone enclosed by 'shells' dominated by bornite, then chalcopyrite, and finally pyrite, which may be up to 15% of the rock. Molybdenite distribution is

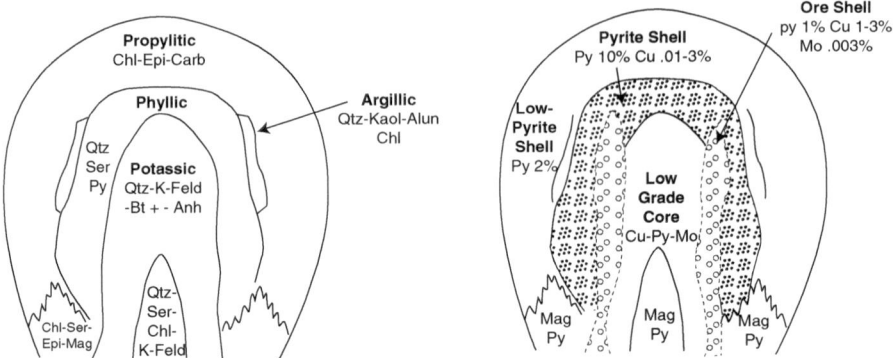

Chl - chlorite ; Epi - epidote ; Carb - carbonate ; Qtz - quartz ; Ser - sericite ; K-Feld - Potassium
Feldspar ; Bt - biotite; Anh - anhydrite ; Py - pyrite ; Kaol - kaolinite ; Alun - alunite ;
Mag - magnetite ; Cu - copper ; Mo - molybdenite

Fig. 4.8 Distribution of alteration zones (*left*) and types of sulfide mineralization in a porphyry
copper deposit (Modified from Lowell and Gilbert 1970)

variable. Radial fracture zones outside the pyrite halo may contain lead-zinc veins
with significant gold and silver contents.

A supergene enrichment zone developed extensively in upper parts of some
deposits. This zone is divided into the oxidized subzone containing unusual
minerals such as chrysocolla, atacamite, antlerite, brochantite, and tenorite with
lesser amounts of malachite and azurite, and the sulfurized subzone of chalcocite,
covellite, native copper and cuprite. These minerals, a series of hydrated Cu
silicates, carbonates, sulfates and oxides, have beautiful green or blue colours and
are prized by mineral collectors.

The composition of the intrusion exerts a fundamental control on the metal
content of the deposit. Low-silica, mafic and relatively primitive plutons, ranging
from calc-alkaline diorite and granodiorite to alkalic monzonite in composition, are
associated with porphyry Cu-Au deposits; intermediate to felsic, calc-alkaline
granodiorites and granites are associated with Cu-Mo deposits; and felsic, high
silica, strongly differentiated granites are associated with Mo, W, and Sn deposits.
The oxidation state, reflected by accessory minerals such as magnetite, ilmenite,
pyrite, pyrrhotite, and anhydrite, also influences metal contents: most deposits are
related to oxidized, magnetite-series plutons, but some Sn and Mo deposits are
related to reduced, ilmenite-series plutons.

Distribution and age: Porphyry deposits are predominantly associated with
Mesozoic to Cenozoic orogenic belts in western North and South America, around
the western margin of the Pacific Basin, and in the Tethyan orogenic belt in eastern
Europe and southern Asia. Major deposits also occur within Paleozoic orogens in
Central Asia and eastern North America and, to a lesser extent, within Precambrian
terranes. Porphyry Cu deposits typically occur in the root zones of andesitic
stratovolcanoes in subduction-related, continental and island-arc settings.

Their distribution can be related to regional structures such as lithosphere-scale faults and rift systems. Cross structures then control the distribution of individual deposits.

Porphyry deposits range in age from Archean to Recent, although most are Jurassic or younger. On a global basis, the peak periods for development of porphyry deposits are Jurassic, Cretaceous, Eocene, and Miocene in age. The youngest deposits are in islands of the southwest Pacific in regions of very active tectonics. The late Miocene Grasberg deposit in Arian Jaya has formed in a zone of intense volcanism and rapid uplift, and will be will be totally removed by erosion within a few million years. This example illustrates the ephemeral nature of this type of deposit and explains why they are typically restricted to young mountain belts.

Origin: As mentioned at the start of the chapter, the close spatial and temporal association between of the ore bodies with granitic intrusions leaves little doubt that magmas are directly linked with the ore-forming process. In fact there are various strong lines of evidence that suggest that both the ore metals and the hydrothermal fluids are derived in large part from the granitic magmas. In addition to the geological aspects, the relationship between metal ratios and magma type, evidence of very high temperatures and the isotopic compositions of the fluids all point in this direction. However, as shown in Fig. 4.9, the compositions of these fluids, as

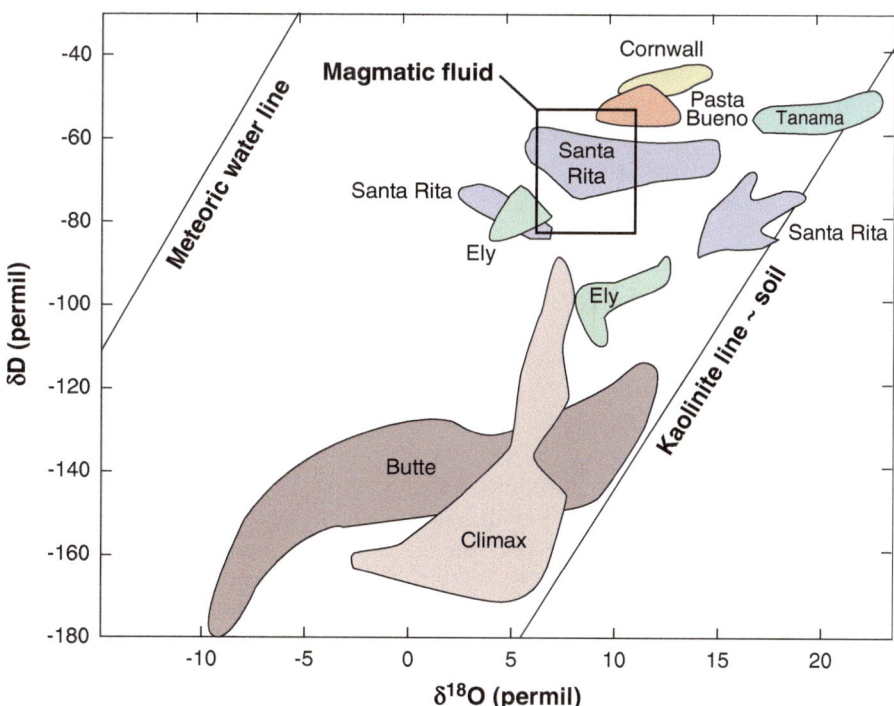

Fig. 4.9 Isotopic composition of oxygen and hydrogen in fluids associated with porphyry deposits (Modified from Barnes 1979)

Box 4.3 The Long History of a Famous Porphyry Deposit, Chuquicamata, Chile

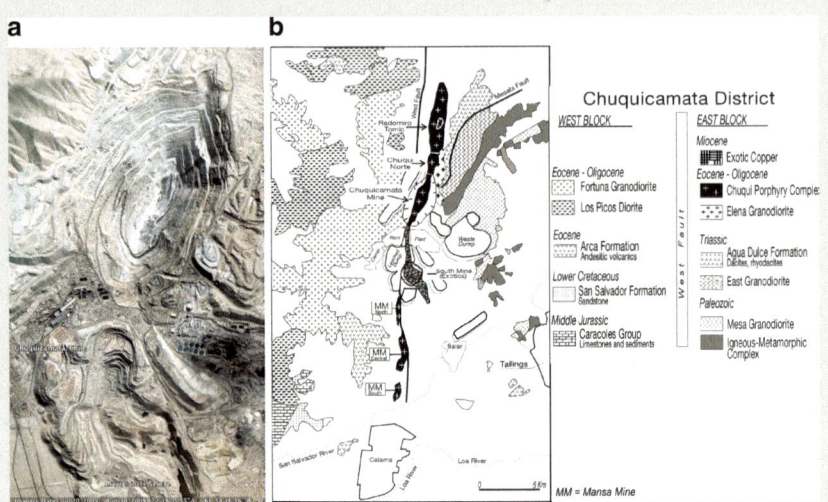

(**a**) Google Earth image of the main Chiquicamata open pit and the smaller Exotica deposit to the south; (**b**) geological map of the deposit (Ossandon et al. 2001).

The Chiquicamata Cu-Mo porphyry mine in the Atacama Desert of northern Chile has been described as the world's greatest mine. With an annual production of copper close to 600,000 t, it was for many years the greatest producer of the metal and despite almost a decade of continual production it still constitutes one of the largest copper resources. It is also a major producer of Mo.

The discovery in 1899 of "Copper Man", a mummy trapped in an ancient mine shaft and dated at about 550 A.D. reminds us that copper has been mined in the region for many centuries. It is claimed that the conquistador Pedro de Valdivia obtained copper for horseshoes from the natives when he passed through in the early sixteenth century.

Mining was limited until the War of the Pacific when Chile annexed large parts of Peru and Bolivia. 'Red Gold Fever' (La Fiebre del Oro Rojo) then drew numerous miners to the Chuquicamata region.

At the beginning of the twentieth century only high grade veins containing 10–15% copper were mined and the disseminated ore was ignored. An attempt in 1899 to process the low-grade ore failed and mining never really developed because of the lack of water, poor communications, a lack of capital, and an unstable copper price.

In 1910, Bradley, an American engineer, finally developed a method of working low-grade oxidised copper ores. He contacted Burrage, a lawyer and industrialist, who approached the Guggenheim Brothers to finance the project. Initial reserves were estimated at 690 million tonnes grading 2.58% copper. The Guggenheims had also developed a process for extracting copper from low-grade ores and in 1912 organised the Chile Exploration Company (Chilex) to mine the deposit. Chilex purchased heavy equipment such as steam shovels (imported from the Panama Canal) and helped build the port at Tocopilla and a 90-mile aqueduct to bring water in from the Andes.

Production started in 1915 and reached 135,890 t in 1929, the year of the Great Depression when demand fell. Companies owned by Guggenheim Bros ran the mine until 1971 when Salvador Allende government nationalized the Chilean copper industry. Since then, Codelco (Corporación Nacional del Cobre de Chile) has mined the deposit.

sampled in fluid inclusions in quartz and other gangue minerals, extend from the magmatic field well into the field of meteoric fluids indicating that the latter are also involved in the ore-forming process.

Putting this all together leads to the following model.

1. A granitic magma is emplaced as a series of pulses into a magma chamber high in the crust, beneath a volcanic edifice. Each pulse cools and partially crystallizes, and as it does, a hydrous fluid phase separates from the silicate magma. The separation of this phase results from one or both of the following processes; (a) the drop in pressure attendant on ascent of the magma decreases the solubility of water in the magma and (b) crystallization of the magma as heat is lost to the wall rocks causes the water content of the residual liquid to build up until it eventually exceeds the solubility limit. Escape of fluid increases the liquidus of the granitic magma, causing the remaining liquid to crystallize rapidly around already crystallized minerals, creating the porphyritic texture characteristic of these deposits. The fluid phase may also migrate up through the silicate liquid, to concentrate at the upper part of the intrusion.

2. The fluid escapes from the inner still liquid interior and moves through fractures in the surrounding solidified carapace and onwards into the wall rocks. As it does so it cools, and it reacts with the wall rocks to form the characteristic alteration that surround all porphyry deposits. The ore metals are transported in the fluids, most probably as chloride or sulfate complexes; as the fluid cools and as its composition changes through reaction with the wall rocks, the stability of the complexes decreases. The ore metals are then precipitated in fractures and within the alteration zones surrounding the granitic intrusion.

3. As the magma intrudes it heats up groundwater in the surrounding rocks, setting up convection cells surrounding the intrusion. The heated groundwater mixes with and reacts with the magmatic fluids, diluting and cooling them and accelerating the precipitation of ore minerals.

4. The process may be repeated several times as new pulses of magma enter the high-level magma chamber, creating a complex, multiphase system of intrusions and ore bodies.

Analysis: Porphyry Deposits
Source of metals – mainly the granitic magma
   Source of S – mainly magmatic
   Source of fluid – magmatic and ground water
   Cause of fluid circulation – expulsion of fluid from the magma, convection of heated groundwater
   Precipitation process – cooling, change in fluid composition, mixing with other fluids

4.3.3   Sedimentary Exhalative (SEDEX) Deposits

The precipitation of sulfides from black smokers is not the only occasion where we can observe a process that forms an ore body. Almost 100 million tones of sediment containing 2% Zn, 0.5% Cu and significant amounts of the Au and Ag has precipitated from hot dense brine that accumulated in the "Atlantis II Deep", a 10 km diameter depression on the floor of the Red Sea. Were this deposit on land and in a politically stable part of the world, it would constitute a very attractive ore body of the type we refer to as a SEDEX or sedimentary exhalative deposit.

Another example is the Salton Sea, a large shallow lake in southern California that formed in 1905 when a canal transporting water from the Colorado River breached and flooded a saltpan. The water became brackish as it dissolved the salt, and large-scale hydrothermal circulation was set up as water in the underlying sedimentary basin was heated by the high prevailing geothermal gradient and the conduits of local active volcanoes. At depth the circulating fluid, a hot (up to 350°C) dense Na-Ca-K-Cl brine, has dissolved Fe, Mn, Pb, Zn, and Cu from the lacustrine sediments that underlie the lake. When the fluid mixes with cooler, dilute surface waters about 100 m below the surface, it precipitates these metals in veins of sulfide. The two processes recorded in Red Sea and Salton Sea examples – precipitation of sulfide-rich chemical sediment and interaction of sediments with circulating hydrothermal fluids – are key elements to the formation of SEDEX deposits.

The definition of these deposits is not straightforward because in many respects they form a continuum with VMS deposits. They typically occur as tabular bodies composed predominantly of Zn and Pb sulfides (sphalerite and galena) and they usually contain economically important amounts of Ag. The Zn and Pb sulfides are interbedded with iron sulfides (pyrite and pyrrhotite) and with generally fine-grained detrital or chemical sediments. They are believed to have formed from hydrothermal fluids that were expelled from mostly reduced sedimentary basins in continental rifts. Two important subtypes are the "Broken Hill type", which is associated with bimodal volcanic rocks and Fe- or Mn-rich chemical sediments, and

Table 4.3 Size and grades of selected SEDEX deposits

Deposit name	Location	Age	Geological resources (maximum size)						
			Cu (%)	Zn (%)	Pb (%)	Ag (g/t)	Au (g/t)	Mt of Ore	Zn + Pb Mt
Broken Hill	Australia	Paleoproterozoic	0.1	11	10	180	0.10	205	43
McArthur River	Australia	Paleoproterozoic	0.2	9.2	4.1	41		237	31
Mount Isa	Australia	Paleoproterozoic		6.8	5.9	148		124	15
Red Dog	U.S.	Mississippian		16.6	4.6	83		165	35
Mehdiabad	Iran	Cretaceous		7.2	2.3	51		218	21
Sullivan	Canada	Mesoproterozoic		5.9	6.1	67		162	19
Navan	Ireland	Mississippian		8.0	2.7			78	8.3
Meggen	Germany	Middle Devonian	0.2	5.8	0.8			60	4.0

"Irish-type" deposits which are hosted predominantly by carbonate rocks. SEDEX deposits comprise 50% of the world's zinc and lead reserves, and 25% of world zinc and lead production.

The general characteristics of selected deposits are given in Table 4.3. Most of these, including the three large Australian deposits and the Sullivan deposit in

Box 4.4 Proterozoic SEDEX Deposits of Northern Australia

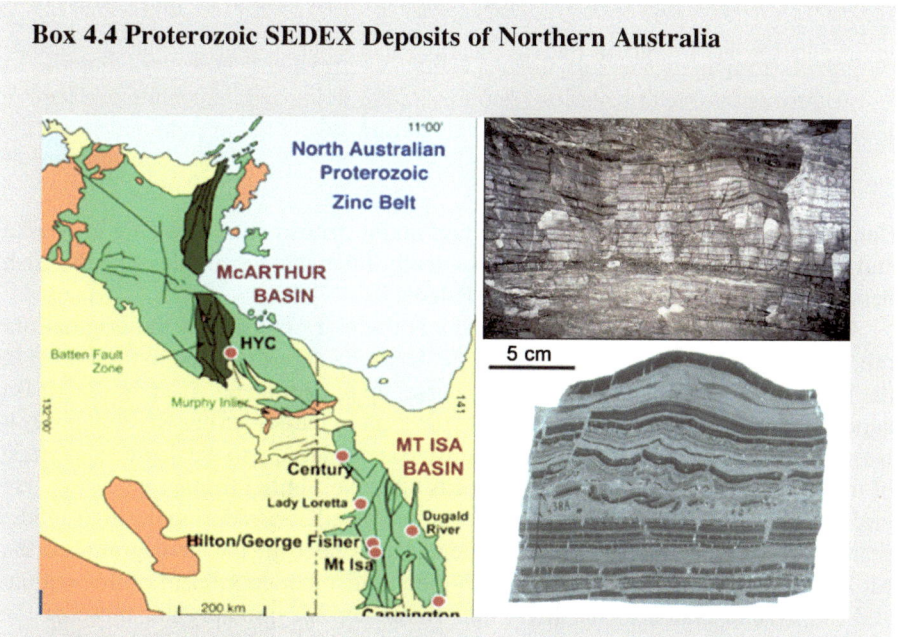

Geological map showing the geology and major SEDEX Pb-Zn deposits in northern Australia. Above right: laminated high-grade Pb-Zn ore in an underground exposure in the HYC mine. Below right: laminated, slightly metamorphosed and deformed ore from the Mt Isa mine (Photos of R. Large)

(continued)

Six large SEDEX deposits are located in two Proterozoic basins in northern Australia. They are very large (14–150 mt) and contain very high grades (average 16% Pb + Zn with significant Ag contents). The Mt Isa deposit was discovered in 1923 and mining started 8 years later leading to a major town in a remote semi-arid part of Australia. The HYC deposit was discovered 30 years later, in 1955, in the more northerly McArthur basin. The name HYC comes from a remark made at the time of discovery: realizing that they had found a major deposit, one geologist turned to the other and said "you have always wanted to name a mine; Here's Your Chance".

The deposits are stratiform and located in intracontinental rifted basins commonly adjacent to major syn-sedimentary faults. The host rocks are dolomitic siltstones and shelf carbonates and the ores are finely laminated, as shown in photos (see also Fig. 1.6). The ore deposits all formed within a relatively short period between 1650 and 1575 Ma.

Opinions differ about the origin of the deposits. The fine banding is very similar to sedimentary bedding and this, together with other features of the ores and their geological setting led Stanton, an influential Australian geologist, to propose in the 1960s an "exhalative-sedimentary" origin, challenging the prevailing view that such deposits were epigenetic and related to granitic intrusions. Other geologists question this interpretation and argue that the textures, mineralogy and chemical compositions of the ores point to their having formed through replacement of pyritic sediments by metals precipitated from hydrothermal fluids. There is agreement, however, that the two processes – sedimentation from exhalative fluids and diagenetic replacement – took place at the sea floor or at shallow levels in the sediment pile.

Canada, are Proteroic but others, including major deposits like Red Dog in Alaska and Mehdiabad in Iran, as well as geologically interesting examples like the Irish deposits and Meggen in Germany, are Paleozoic.

A characteristic feature of the deposits is the fine grainsize of the original ore minerals. Compare, for example, the three Australia deposits listed in the Table: in the Broken Hill orebody the grains are coarse, up to centimeter sized, in the Mt Isa deposit the average grainsize is 100–500 μm, and in the McArthur River deposit, it is less than 10 μm. The difference in grainsize is due largely to the degree and grade of metamorphism that affected the deposits after their initial formation. The coarse grains of the Broken Hill deposit result from recrystallization during the high-grade, granulite-facies metamorphism that affected this deposit; the finer grains of the Mt Isa deposit are influenced by the sub-greenschist metamorphism of this region; and the minute grains of the McArthur River deposit probably are those of the original sedimentary ore minerals (This difference in grainsize strongly influences the viability of the three deposits. As mentioned in Chap. 2, the coarse Broken Hill ores are easily mined and refined, in contrast to the ultra-fine McArthur River ores, which for many years were unable to be exploited).

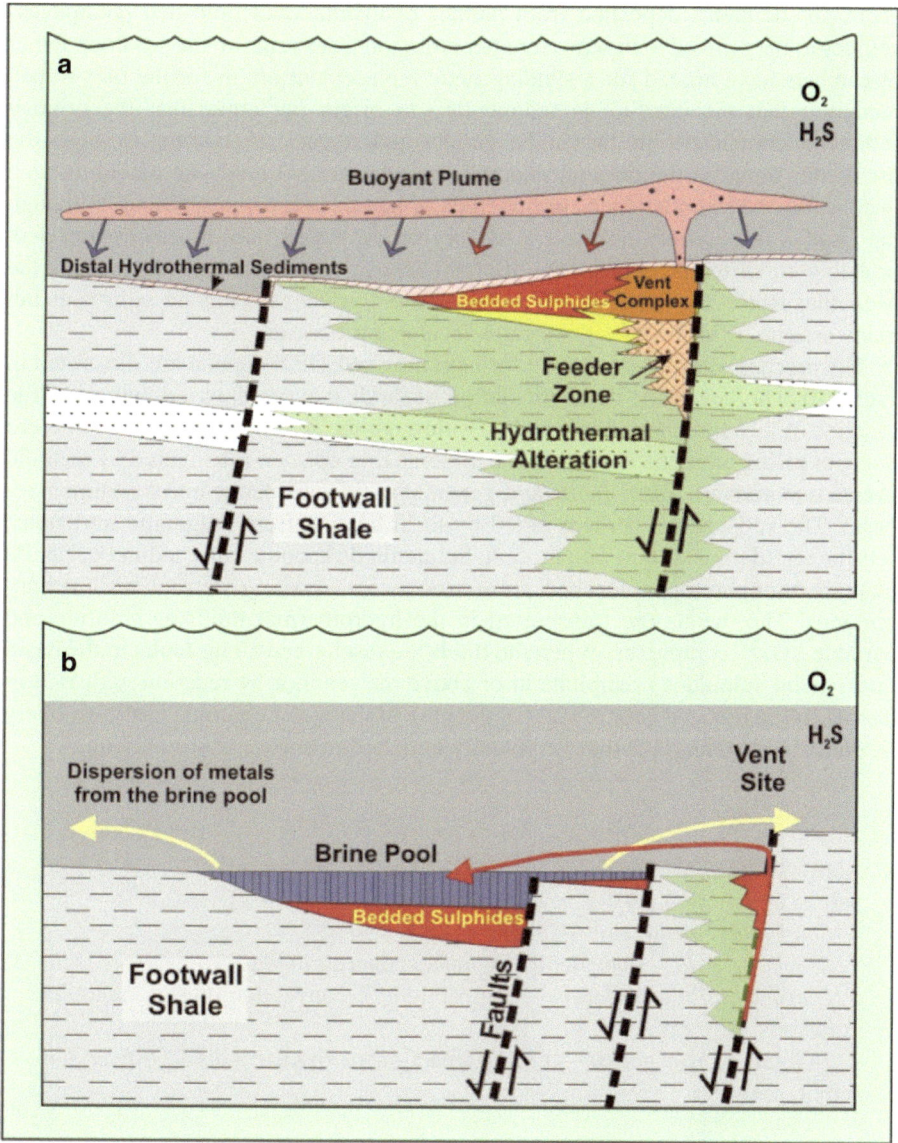

Fig. 4.10 Illustration of the origins of proximal and distal SEDEX deposits (Modified from Goodfellow and Lydon 2007)

Another striking feature of a SEDEX deposit is the banding displayed in many ores. In the samples shown in Fig. 1.6a and b the alternation of bands of light-coloured sulfides and darker silicates is clearly visible. The form and structure of the bands strongly resembles fine sedimentary bedding, as in chemical sediments such as cherts or banded iron formations, and this resemblance has led many workers to interpret the ores as sedimentary-exhalative in origin: i.e. they are interpreted as chemical sediments deposited from plumes of hydothermal fluid that precipitated sulfides as they mixed with seawater after emission from vents on the sea floor. Other researchers have argued for a syndiagenetic replacement origin for the ores – they recognise that the banding is sedimentary in origin but argue that the original sediments contained only barren Fe sulfides and silicates. According to them, the ore metals replaced the original minerals as hydrothermal fluids circulated through unconsolidated sediments tens or hundreds of metres below the sea floor. Although the issue is not entirely resolved, it is very probable that both processes operated, probably to different extents in different deposits. The Zn and Pb ores of the McArthur River, for example, do seem to have formed as chemical sediments but many aspects of Mt Isa ores point to replacement processes.

The processes implicated in the formation of a SEDEX deposit are illustrated in Fig. 4.10. The key is the deep circulation of fluids that are drawn down along the margins of a sedimentary basin and pass through the sedimentary sequence before being expelled on to the sea floor. The mineralizing episode is triggered by tectonic events that activate major faults and generate rapid subsidence in the sedimentary basin. The subsidence, perhaps aided by local heating from magmatic intrusions, sets the circulating system into motion. Saline fluids become enriched in Fe, Zn, Pb that are thought to be leached from iron oxides coating detrital sedimentary minerals. The metals are transported in the hydrothermal fluids as chloride and variable SO_4^{2-} complexes. When the fluids are discharged along faults to the basin floor, metal sulphides precipitate at or above the seafloor by reaction with H_2S in the overlying reduced anoxic layer at the base of the water column. The most likely S source is biogenic H_2S that is typically enriched in anoxic water columns.

Analysis: SEDEX Deposit
Source of metals – detrital sedimentary rocks
   Source of S – biogenic H_2S
   Source of fluid – seawater and connate (interpore) water
   Cause of fluid circulation – compaction(?), convection due to magmatic intrusions
   Precipitation process – cooling, reaction of oxidised fluid with H_2S in anoxic seawater

4.3.4 Mississippi Valley Type (MVT) Deposits

This type of deposit is the antithesis of porphyry deposits: they form at very low temperatures and they have nothing whatsoever to do with magmas. The name comes from the valley of the Mississippi River in central USA where these deposits were first recognized. They form a varied family of epigenetic lead-zinc ore deposits that occur predominantly in carbonates of Paleozoic (Cambrian to

a

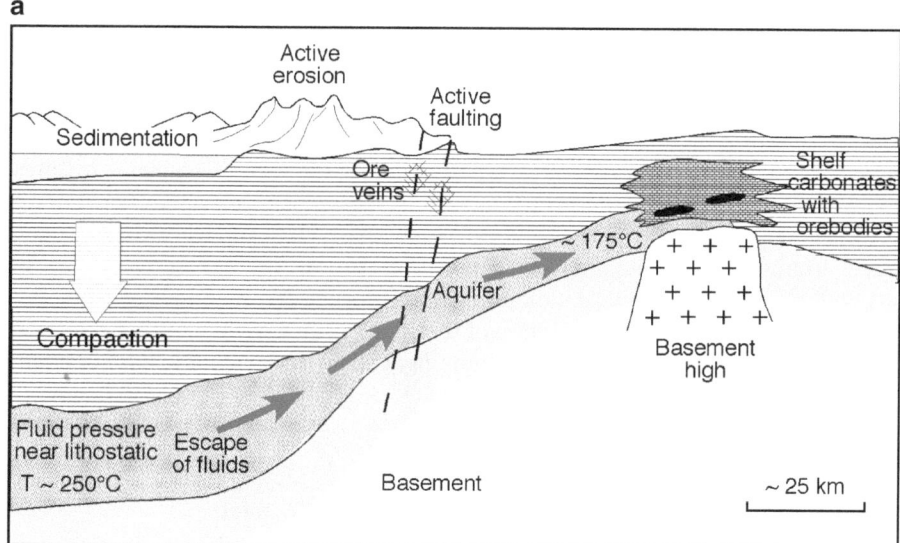

b

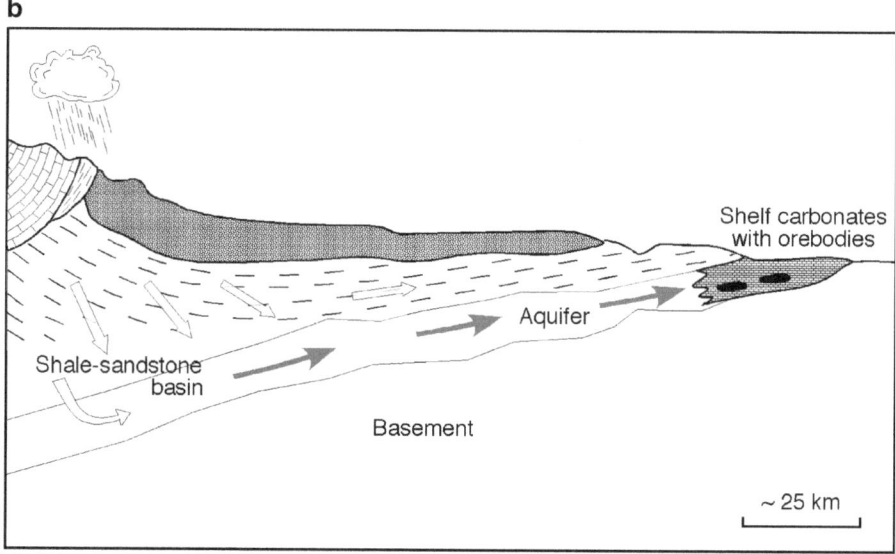

Fig. 4.11 Illustration of the origin of MVT deposits (From Evans 1993)

Triassic) ages. MVT deposits of Cretaceous age are found in Algeria and Tunisia
but examples in Precambrian rocks are very rare.

MVT deposits are epigenetic and stratabound and occur in dolostones, or less
commonly in limestone or sandstone, at shallow depths at flanks of sedimentary
basins. A common depositional setting is in platform carbonate sequences, commonly reef facies, located either in relatively undeformed foredeeps or in foreland
thrust belts. (Fig. 4.11). Most deposits constitute parts of ore "districts" that cover

many hundreds of square kilometers and contain numerous small to large deposits. The limits of an ore district are defined by geologic features, most notably the presence of breccias, facies changes from shale to carbonate at basin margins, large faults and basement highs.

The deposits are mineralogically simple; dominant minerals are sphalerite, galena, pyrite, marcasite, dolomite, calcite, and quartz. Sulfide mineral textures are extremely varied, ranging from coarse and crystalline to fine-grained, massive to disseminated. Banded and colloform structures typical of deposition in open spaces from fluids are found in some deposits (Box 4.5). Alteration associated with ore bodies consists mainly of dolomitization, brecciation, host-rock dissolution, and the dissolution or recrystallization of feldspar and clay. Evidence of dissolution of carbonate host rocks, expressed as slumping, collapse or brecciation, is common.

Box 4.5 The Missouri-Mississippi Valley Mining District

(a) Brecciated ore from the Robb Lake MVT deposit in Canada. A matrix of sphalerite and galena encloses fragments of dolomite. (b) Colliform ore from the Cadjibut mine in Australia (Photo – Chris Arndt).

During the last centuries, the mining industry has been important for the economic and social fabric of several states of central USA and especially for Missouri. Pierre Charles LeSeur, a Frenchman, first prospected in the

Mississippi Valley in the beginning of the eighteenth century. He found plentiful shiny gray mineral (galena) at the surface and since this discovery, Missouri has been the major source of lead of USA. The metal originally was used as a roofing material.

The southeastern Missouri Mississippi Valley-type Mineral District contains some of the highest concentration of lead on the world as well as large quantities of zinc, copper and silver. The ore are primarily hosted by bacterial stromatolite reefs and associated oolitic rocks of a Cambrian dolomitic formation deposited in a shallow sea. They formed when warm metal and organic-carbon bearing fluids migrated from adjacent sedimentary basins through this formation.

For many years deposits Mississippi Valley deposits, located in relatively young (Paleozoic) sedimentary basins, provided almost all the lead consumed in the USA; Europe, in contrast, was supplied by large Australian deposits such as Broken Hill and Mt Isa, which are located in rocks of Proterozoic age. (The latter, both SEDEX deposits, were described in the previous section). The difference in age is transmitted to the isotopic composition of the lead, which is considerably more radiogenic in the case of the older Australian deposits. Particularly during the period 1950–1990, before lead was withdrawn from petrol, this isotopic difference was used as a tracer of pollution. Dust and other pollutants blown eastward across the Atlantic could easily be distinguished from material from local European industry by the isotopic composition of its lead. And in archeological studies, lead from local sources mined from Roman to modern times could readily be distinguished from modern industrial sources.

Ore deposition temperatures determined from fluid inclusion studies are low (50–200°C), but somewhat higher than those attributable to normal thermal gradients within the sedimentary pile. Ore fluids were dense basinal brines, typically containing 10–30 wt.% dissolved salts. Lead and sulfur isotopic data indicate that the sources for both metal and reduced sulfur were the sedimentary rocks of the basin.

Within each ore district, deposits display remarkably similar features, including mineral assemblages, isotopic compositions, and textures. Ore controls typically are district-specific; examples include shale edges (depositional margins of shale units), limestone-dolostone transitions, reef complexes, solution collapse breccias, faults, and basement topography. Most MVT ore districts are the product of regional or sub-continental scale hydrological processes. Therefore, diversity among MVT districts is expected because of wide ranging fluid compositions, geological and geochemical conditions, fluid pathways, and precipitation mechanisms possible at the scale of MVT fluid migration.

Origin: As with many other types of deposits, the broad outline of the ore-forming process is well understood but the details, some of crucial importance, remain obscure. As mentioned above, there is strong geological and geochemical evidence that both the metals and the sulfide now found in the ore bodies were derived from the detrital sedimentary rocks of the sedimentary basin. There is equally strong evidence that the hydrothermal fluid was connate water; i.e. the fluid, initially seawater, that filled the pore-space between the detrital grains of the poorly consolidated sedimentary rocks. Even the mechanism that triggers the precipitation of ore sulfides is well understood. Basinal fluids are relatively oxidized and they most probably transported the metals as chloride or sulfate complexes. The geochemical environment of the carbonates that now host the ores was very different – it was reduced and contained abundant reductants in the form of hydrocarbons (oil or gas) and other organic material. The redox reaction destabilized and reduced the chloride or sulfate complexes, causing the precipitation of Pb and Zn sulfides. Platform carbonates are often highly porous, due to the presence of the breccias and cavities that develop during dolomitization, and the ores were precipitated in these cavities or in zones of reaction between the fluid and the carbonate rocks.

What is unclear is the process that sets the fluid in motion – stagnant connate water cannot form a large ore body; to do so, the fluid must migrate to the margins of the basin in order that it can interact with the reductants in the carbonates. Various processes are debated in the literature. Dewatering of the basin during compaction under the load of overlying sediment is commonly advocated, but this process is probably too slow to explain the fluid fluxes inferred for many deposits. A remarkable result of geochronological studies of the ore bodies and their broad-scale geological setting has shown that the timing of ore formation in the Mississippi Valley in central USA coincides with major deformation events in the Appalachian mountain range at the Atlantic margin of the continent. This association led to the idea that thrusting at the eastern margin of the sedimentary basin drove the ore-forming fluids for over 1,000 km until they reached and reacted with the marginal reefs at the other side of the basin (Fig. 4.11).

Detailed recent studies have shown that the ores formed when a pulse of warmer-than-normal fluid was injected into the carbonate platforms. The duration of the pulse was too short to have been formed by slow-acting processes like compaction or orogenic movements and this has led to imaginative alternatives such as the flooding of the basin by warm seawater following changes in the movement of major ocean currents or transgression following melting of continental ice caps.

Analysis: MVT Deposits

Source of metals – detrital sedimentary rocks of the basin;

   Source of S – biogenic H_2S or sedimentary sulfide

   Source of fluid – connate (interpore) water

   Cause of fluid circulation – compaction (?), tectonic deformation (?), sea level increase (?)

   Precipitation process – redox reactions as oxidized basin water meets organic material in carbonate facies

4.4   Other Types of Hydrothermal Deposit

The deposits described above we selected in part because they are economically important in that they provide most of the world's supply of base metals, and also because they illustrate the wide range of processes that are involved in the formation of a hydrothermal ore body. The list is by no means complete, however, and in the following section we provide brief descriptions of some other types.

4.4.1   Stratiform Sediment-Hosted Copper Deposits

These rank just behind porphyry deposits as a source of copper and represent the most important source of Co. Some examples are also rich in Pb, Zn, Ag, U, and Au.

Stratiform sediment-hosted copper deposits are found in intracontinental rift-related sedimentary sequences and typically at junctions between oxidised Aeolian

> **Box 4.6 Kupferschiefer in Central Europe and the Central African Copperbelt**
>
> The two main locations stratiform sediment-hosted copper deposits are in the Permo-Triassic Kupferschiefer ("copper shale") of Germany and Poland and the Central African Copperbelt. The first is well known in a historical context. The deposits of the Kupferschiefer have been mined more or less continuously since the middle ages and in the sixteenth century Georgius Agricola, the first mineralogist, laid the foundation for the systematic and scientific study of geology and mining. His remarkable book *De Re Metallica*, published in 1556, describes miners and mining of deposits in the Kupferschiefer; he notes, for example, a spatial relationship between bituminous shales and the copper mineralization, anticipating, by over 500 years, modern ideas of ore formation.
>
> The history of exploitation of the deposits of the Copperbelt in Zambia (initially Northern Rhodesia) and the Congo provides some interesting, and troubling, lessons. During much of the twentieth century the deposits were the backbone of the economies of what were then British and Belgium colonies. Up until the 1970s the mines were run efficiently (though most of the wealth went to the colonial rulers) and their presence fuelled economic hopes for the post-colonization period. Their importance was severely diminished, however, by a crash in global copper prices in 1973, compounded by the nationalization of the copper mines by the governments of the newly independent nation. During the following 30 years production in these enormous and rich deposits fell almost to zero as a result of corruption, neglect and mismanagement, and only at the turn of the century has mining revived. The period 2007–2009 has seen an influx of new investment from Chinese government agencies, initiating, perhaps, a new period of economic colonization.

sandstones and more reduced assemblages of shales, carbonates and evaporates. The metals were transported in basin-derived fluids that were set into motion by rapid rifting and subsidence. The metals were leached from detrital minerals such as magnetite, biotite and hornblende and they were transported as chloride complexes. Ore deposition occurred at the redox interface between oxidised and reduced sedimentary rocks. As for SEDEX deposits, there is considerable debate about the precise process, and particularly whether the ore metals were primary precipitates or epigenetically replaced sedimentary iron sulfides.

4.4.2   Uranium Deposits

Uranium is very different from the other elements discussed in this chapter: it is an energy source, and not a metal used in industry or finance like the copper, zinc, or gold; and because it is radioactive, used in bombs, it is the target of the ire of ecologists (a moustachioed politician who became famous for tearing down a MacDonald's and brandishing his roquefort at anti-capitalism demonstrations, learnt his trade in anti-nuclear protests). Although a trace element, uranium is found in a large range of crustal rocks and forms a wide variety of deposits. A brief description of the more important types in given in Table 4.4. Those in magmatic rocks and in purely sedimentary settings are mentioned in other chapters; here we discuss just two types, unconformity-related deposits and sandstone deposits, both of which formed from hydrothermal fluids, to continue the theme of ore deposition related to redox reactions.

The primary uranium ore mineral in these and other deposits is uraninite (UO_2) or pitchblende (UO_3, U_2O_5). Other uranium minerals include carnotite ($K_2(UO_2)_2(VO_4)_2 \cdot 3H_2O$) and complex oxides or titanates rich in rare trace elements such as davidite-brannerite-absite, and the euxenite-fergusonite-samarskite group. Secondary uranium minerals such as torbernite and autunite have brilliant yellow or green colours and are fluorescent under ultraviolet light.

The key to the formation of uranium deposits is the vastly different solubility of this element in oxidized and reduced fluids. Uranium occurs in two valence states, the reduced form U^{4+} and the oxidised form U^{6+}. The latter is highly soluble in oxidised fluids where it forms stable complexes with fluoride, phosphate or carbonate ligands; in this condition it is readily transported in the fluids that circulate in sedimentary basins. The reduced form, in contrast, is highly insoluble, such that when an oxidised fluid comes into contact with a reductant, the U precipitates.

The richest uranium ore bodies are the *unconformity-related deposits* in the Athabasca Basin, in Saskatchewan, Canada. These deposits are not large, almost always less than one million tons of ore, but their relatively small size is compensated by high grade; Cigar Lake contains about 875 000 t of ore at an average grade of 19% uranium oxide and McArthur River a slightly smaller amount at an average grade of 24%. Similar deposits in the Northern Territories of Australia are larger but have far lower grade, averaging 0.4%.

Table 4.4 Summary of characteristics of main types of uranium deposits

	Age	Grade	Proportion[a] (%)	Geological features	Examples
Unconformity-related deposits	Proterozoic	0.4–24%	30	Near major unconformities between sandstones at the base of sedimentary basins and metamorphic basement rocks	Athabasca Basin, Canada; McArthur Basin, Australia.
Sandstone deposits	Paleozoic-Cenozoic	0.05–0.4%	15	Sandstones in a continental fluvial or marginal marine sedimentary environment interbedded with shale or mudstone	Wyoming Basin and Colorado Plateau, USA; Central Europe, Kazakstan
Quartz-pebble conglomerate deposits	Paleoproterozoic	0.01–0.15%	10	Stratiform and stratabound paleoplacer deposits	Witwatersrand, South Africa; Elliot Lake, Canada
Iron-oxide copper gold deposits	Proterozoic	0.04–0.08	40	Hematite-rich granite breccia	Olympic Dam, Australia
Intrusion associated deposits	Proterozoic	0.03	5	Veins in leucogranite	Rossing, Namibia
Volcanic deposits	Precambrian to Cenozoic	0.02–0.2%	<1	Veins and breccias in felsic to intermediate volcanic rocks	Streltsovskoye, Russia; Dornod, Mongolia; McDermitt, Nevada.
Surficial deposits (calcretes)	Tertirary to recent	0.15	5	Near-surface uranium concentrations in sediments or soils	Yeelirrie, Australia

[a] Approximate percentage of global uranium resources

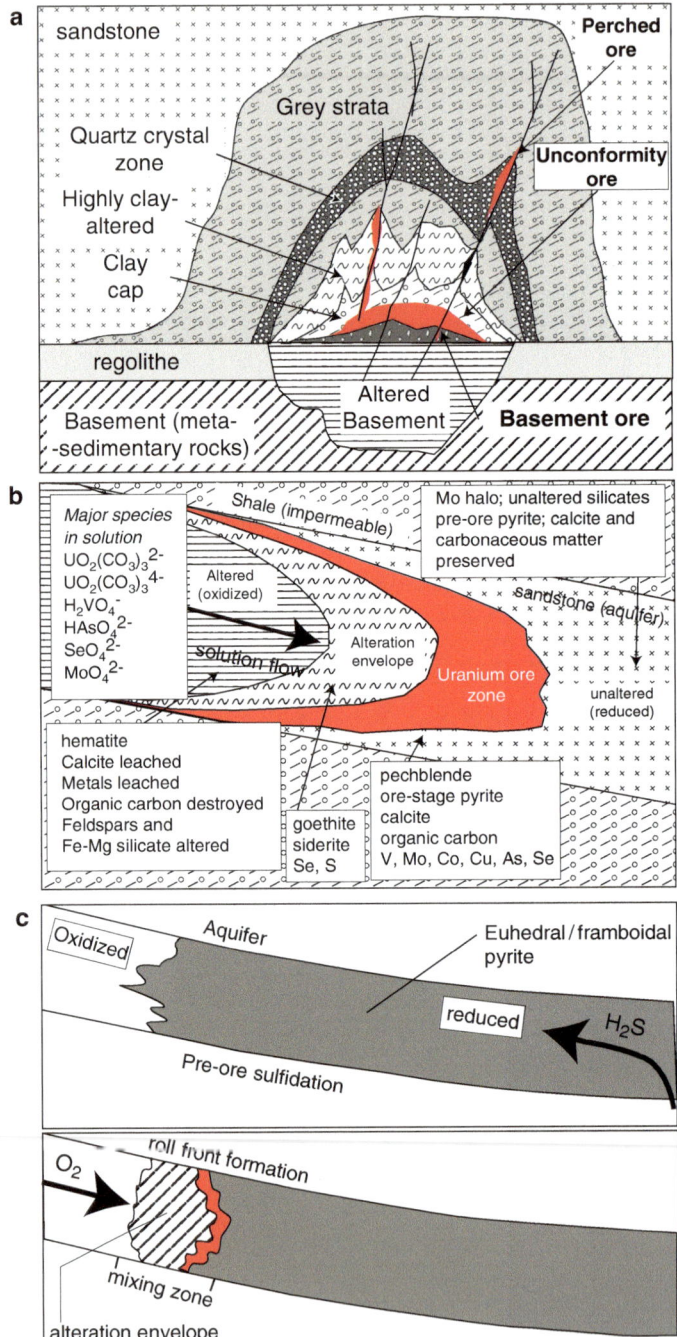

Fig. 4.12 Illustration of the geology and origin of uranium deposits (**a**) roll-front, (**b**) unconformity-related deposits (Modified from Jefferson et al. 2008 and Robb 2005)

In the *unconformity-related deposits* deposits, pods, veins, and semimassive replacements of uraninite are located close to unconformities between early Prote-rozoic conglomeratic sandstones in the lower portions of intracratonic basins and metamorphosed basement rocks (Fig. 4.12a). Fluids in the sandstones are oxidized and as they circulate they dissolve uranium from detrital minerals such as monazite, alanite and apatite, which were derived initially from granites of the basement. When these fluids come into contact with pockets of pelitic, organic-rich schist in the basement, the uranium precipitates to form an ore body. The exact location of the ore body is strongly influenced by variations in the permeability of the sandstones and the presence of faults that control the fluid circulation.

Sandstone deposits are the principal source of ore in the USA, mainly from deposits of the Wyoming Basin and Colorado Plateau. These deposits are contained within medium to coarse-grained sandstones deposited in a continental fluvial or shallow marine sedimentary environment. There are two main types, referred to as "tabular" and "roll-front" deposits. In the former, impermeable shale or mudstone units are interbedded in the sedimentary sequence and occur immediately above and below the mineralized horizon. The fluid within the sandstone is low-temperature, low-salinity, oxidized meteoric water that flows readily through the permeable sandstone, transporting with it dissolved uranium; the fluid in the shales and mudstones is a relatively stagnant, salt-rich basinal brine. Mixing between the two fluids at the interface between the sedimentary units precipitates uranium minerals.

"Roll-front" deposits form in a similar geological setting but via a different mechanism. The host rock is permeable sandstone which at depth contains a reduced assemblage of pyrite, calcite and organic matter. Oxidized fluid flowing from nearer the surface down the permeable horizon reacts with the reduced material creating at redox front – a cusp-shaped zone where the two types of fluid mix and react (Fig. 4.12b). The uranium minerals initially precipitate at the front, which continues to migrate down and along the sandstone layer. As it moves it sweeps up the dispersed uranium in the reduced material concentrating it at the redox front, thus creating a richer and richer deposit.

4.4.3   Iron-Oxide Copper Gold (IOCG) Deposits

An IOGC deposit is defined as a polymetallic, breccia-hosted deposit in which ore is spatially and temporally associated with granite and with iron oxide alteration. The Olympic Dam deposit is located within a funnel-shaped, hematite-rich hydro-thermal breccia that formed close to the surface through progressive, polyphase fracturing and alteration of the upper part of a granitic intrusion. The ore minerals are diverse and complicated, comprising some 30 varieties of Cu, U, Au, Ag, Ni, Co sulfides, sulfosalts, oxides, carbonates, and native metals. They occur as veins, disseminations, irregular patches and breccia fillings that occur together with zones of intense calcic-sodic, iron and potassic alteration.

Box 4.7 Discovery of the World's Biggest Mine

In 1975 geologists and geophysicists from Western Mining Corporation, then a medium-sized Australian mineral-exploration and mining company, were exploring for copper deposits in the Gawler Block of South Australia. The target was elusive, in the desert, hidden beneath 300 m of younger sediment formations, and 100 km away from the closest known mineralization. Doug Haynes, a geologist just out of his PhD, had developed the idea that copper deposits might form from a basaltic source via the oxidation of magnetite. The first hole they drilled intersected a magnetite breccia containing a small concentration of copper – a tantalizing hint – but then holes 2–9 found nothing. Finally persistence (and an unusual level of support from the Melbourne head office) paid off and hole number 10 intersected 200 m of ore containing 2% copper and significant tenors of gold and uranium. The team had discovered one of the richest ore bodies in the world and an entirely new type of ore deposit.

The Olympic Dam deposit contains almost eight billion tons of copper-uranium-gold ore: it is the world's biggest uranium resource, the fifth largest gold deposit, and one of the biggest copper deposits. When expansion planned in coming years is completed, it is expected to become the world's biggest mine. Polymetallic deposits such as Olympic Dam are particularly attractive to mining companies because the prices of metals normally do not vary in unison – the gold price, for example, tends to increase during periods of recession thereby protecting the companies in times of trouble. The discovery of Olympic Dam set off exploration programs for similar deposits throughout the world and led geologists to take a new look at many existing deposits which were subsequently reclassified as iron-oxide copper gold (IOCG) deposits.

(For a more complete account of the Olympic Dam discoverey, see http://www.science.org.au/scientists/interviews/w/woodall.html#9)

Due to the very recent discovery of the deposit type, theories of ore formation are subject to continual revision; most call on large-scale magmatic events that drive large-scale flow of oxidized probably magmatic hydrothermal fluids into mid to upper crustal levels along fault zones. Mixing of these fluids with near surface meteoritic fluids or brines is commonly invoked as the ore-forming process.

4.4.4 Gold Deposits

"Well, have you found any gold?" This question comes up when any geologist or student on a field excursion talks to a local farmer. Even in Europe, a continent with few gold deposits of any size (at least that have so far been discovered), the population expects a geologist to look for gold. This opinion is no doubt coloured by the gold rushes that saw millions of Europeans flock to the New World, first to California, then to Australia, and finally to the Yukon and Alaska between 1850 and 1900.

The gold that was first mined by successful prospectors during the gold rushes came from placer deposits in streams and rivers or desert sands. This type of deposit is discussed in the following chapter. Placer gold is released by weathering, erosion and transport of the metal from primary deposits, almost all of which are originally hydrothermal in origin. As with uranium deposits, the types of fluid that deposit the ores are diverse, but gold-forming fluids tend to be hotter, and they commonly have closer association with magmas than those responsible for most uranium or base metal deposits.

In its natural state gold usually occurs in the native form, as alloys containing upwards of 85% Au with lesser amounts of silver, copper or in some cases platinum-group metals. In some deposits tellurides such as sylvanite ([Au,Ag] Te_2) are important. Quartz is by far the most common gangue mineral accompanied by carbonates (calcite, dolomite, ankerite) and sulfides (pyrite, pyrrhotite, arseno-pyrite, galena). Because of the high price of gold, the metal can be mined even when grades are very low. A high-grade gold deposit contains 10–150 g per tonne (and only minor concentrations of other valuable metals) and in the biggest most economical open-pit mines, the grade may be less than 1 g/t. Gold is also extracted as a by-product in other types of deposits, as in the porphyry deposits described earlier in the chapter or in iron-oxide-copper-gold deposits.

Deposits related to magmatic fluids: One class of gold deposits is given the name "epithermal" with reference to the classification of Lindgren, the great American geologist, who coined this term for deposits that form from hydrothermal fluids at shallow crustal levels. Deposits of this type have been found in increasing numbers in the magmatically active circum-Pacific region. Studies of active hot springs and of fluid inclusions in gangue minerals show that fluids in such regions are hot, 160–270°C, with contrasting pH and oxidation states. Variations in these parameters lead to two types of deposit, called *high-sulfidization* when they are were derived from fluids containing oxidized sulfur species (SO_2, SO_4^{2-}, and HSO_4^-) or *low-sulfidization* when from fluids with reduced sulfur species (H_2S, HS^-). The former type commonly forms close to volcanic vents from fluids derived directly from the magma; the latter in distal parts of the volcanic edifice from mixtures of magmatic and meteoric fluids (Fig. 4.13)

Gold is transported as chloride (Cl^-) or sulfide (HS^-) complexes whose stability depends crucially on the composition, pH and Eh of the fluid. When these parameters change, the complexes break down and the metals come out of solution. In the case of low-sulfidation fluids, gold precipitates when the boiling of the fluid causes loss of H_2S to the vapour phase, or when it mixes with cool, oxidized meteoric water. For high-sulfidation fluids the cause of Au precipitation is less well understood.

Orogenic gold deposits: The second major class of gold deposits consists of quartz or quartz-carbonate veins in deformed and metamorphosed terranes in

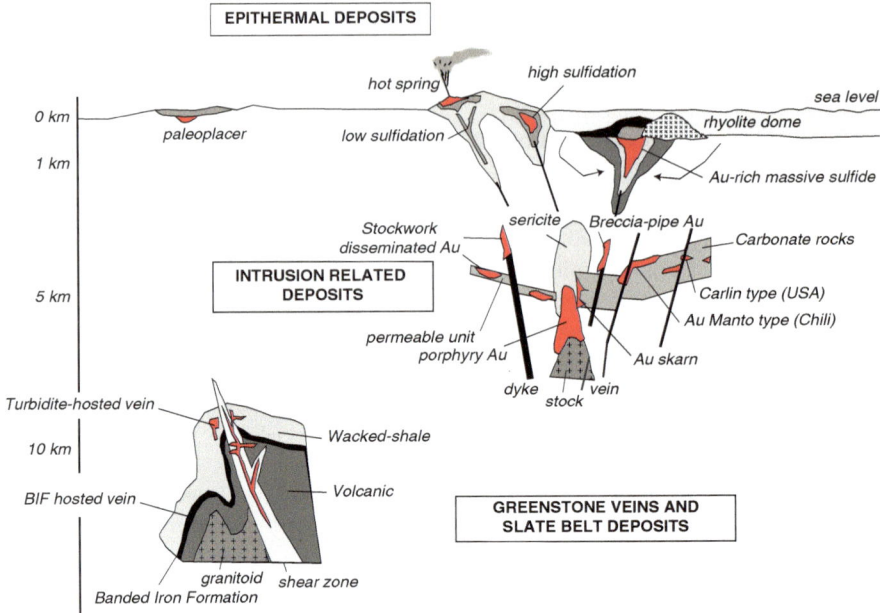

Fig. 4.13 Geological settings of the various types of gold deposits (Modified from Dubé and Gosselin, 2008)

convergent margin settings of all ages. The metamorphic grade is greenschist facies or more rarely amphibolite to granulite. The gold-bearing veins are spatially associated with crust-scale deformation zones and usually show strong local structural control. Hydrothermal alteration surrounding the veins contains the same minerals as in the gangue – quartz, carbonates and sulfides with additional low-temperature silicate minerals such as sericite, albite, biotite, and chlorite. One group of well-known deposits includes examples in Archean greenstone belts such as Kalgoorlie in Australia and Timmins-Kirkland Lake-Val d'Or in Canada. These are the source of about 15% of global gold production. Another group includes the deposits in California and the Klondike in North America and Victoria in Australia. Weathering of these deposits produced the placers that fuelled the gold rushes in these regions, as described in the following chapter.

Orogenic gold deposits have no direct association with magmas. Opinions differ as to the source of the ore fluids: possible origins include (1) metamorphic fluids released by the dehydration that accompanies the breakdown of hydrous minerals during amphibolite to granulite facies metamorphism in the middle to lower crust; (2) magmatic fluids from deep granitic intrusions; (3) CO_2-rich fluids from a subcrustal source; (4) deeply circulating meteoric water. As these fluids are driven through the crustal sequence of metavolcanic and metasedimentary rocks, they leach out gold which is transported, once again, as chloride or sulfide complexes. The fluids are channelled along major structural discontinuities and react with the wall rocks at higher levels in the crust; the reaction produces the distinctive alteration zones that surround the ore veins, changes the compositions of the fluids, and leads to deposition of gold and other minerals. Structure exerts a major control on the location of ore deposits, guiding the passage of ore fluids and influencing the sites of ore deposition. Preferred sites include zones of dilation at the intersections of fault zones or in fold hinges, or zones of brecciated or sheared rock that are contain pore spaces that are filled with secondary minerals or friable or reactive rocks that are replaced by the ore minerals.

Carlin-type deposits: This important class of deposits is the source of most of the gold mined in the USA. The name comes from several large deposits related to the Carlin unconformity in Nevada. The gold, which associated with antimony, mercury and thallium, is very finely disseminated in Paleozoic silty carbonate host rocks. Ore deposition took place in the Eocene from relatively low-temperature (150–250°C), low pH, moderately saline hydrothermal fluids. No intrusions are associated with the deposits and opinion is divided as to whether the fluids are of meteoric, metamorphic or magmatic origin

References

Barnes HL (1979) Geochemistry of hydrothermal ore deposits. Wiley, New York, 997 pp

Brimhall GH, Crerar DA (1987) Ore fluids: magmatic to supergene. Rev Mineral Geochem 17:235–321

Cathles LM, Adams JJ (2005) Fluid flow and petroleum and mineral resources in the upper (<20 km) continental crust. Econ Geol 100th Anniversary Volume:77–110

Chenovoy M, Piboule M (2007) Hydrothermalisme. Spéciation métallique hydrique et systèmes hydrothermaux. Collection Grenoble Sciences 624 pp

Dubé, B. and Gosselin, P. 2008, Mineral Deposits of Canada - greenstone-hosted quartz-carbonate vein deposits. http://gsc.nrcan.gc.ca/mindep/synth_dep/gold/greenstone/index_e.php.

Ellis A.J., (1979) Explored geothermal systems. In H.L. Barnes (Ed.), Geochemistry of hydrothermal ore deposits, 2nd ed. : 632–683. Wiley & Sons

Evans AN (1993) Ore geology and industrial minerals: an introduction. Blackwell, Oxford, 390 pp

Franklin, JM and Gibson, HL and Jonasson, IR and Galley, AG (2005) Volcanogenic massive sulfide deposits. Economic Geology 100th Anniversary Volume, pp 523–560

Goodfellow WD, Lydon JW (2007) Sedimentary exhalative (SEDEX) deposits. In: Goodfellow WD (ed), Mineral deposits of Canada: Geological Association of Canada. Special Publication No. 5, pp 163–183

Hannington MD, Galley AG, Herzig PM, Petersen S (1998) Comparison of the TAG mound and stockwork complex with Cyprustype massive sulfide deposits. In: Proceedings of the Ocean Drilling Program, Scientific Results Volume 158, College Station, pp 389–415

Hedenquist J.W., Henley R.W., (1985) Hydrothermal eruptions in the Waiotapu geothermal system, New Zealand: origin, breccia deposits and effect of precious metal mineralization. Economic Geology (80) : 1640–1666.

Herrington, R., Maslennikov, V., Zaykov, V., Seravkin, I., Kosarev, A., Buschmann, B., Orgeval, J.J., Holland, N., Tesalina, S., Nimis, P., Armstrong, R. (2005) Classification of VMS deposits: Lessons from the South Uralides. Ore Geology Reviews (27), pp203–237. doi:10.1016/j.oregeorev.2005.07.014

Krupp R.E., Seward T.M., (1987) The Rotokawa geothermal system, New Zealand: an active epithermal gold-depositing environment. Economic Geology (82) 1109–1129

Jefferson CW, Thomas DJ, Gandhi SS, Ramaekers P, Delaney G, Brisbin D, Cutts C, Quirts D, Portella P, Olson RA (2008) Unconformity associated uranium deposits. In: Goodfellow WD (ed), Mineral deposits of Canada: Geological Association of Canada. Special Publication No. 5, pp 273–305

Lowell JD, Gilbert JM (1970) Lateral and vertical alteration-mineralization zoning in porphyry ore deposits. Econ Geol 65:373–408

Ossandon CG, Freraut RC, Gustafson LB, Lindsay DD, Zentilli M (2001) Geology of the Chuquicamata mine: a progress report. Econ Geol 96(2):249–270. doi:10.2113/96.2.249 DOI:dx.doi.org

Robb LJ (2005) Introduction to ore-forming processes. Blackwell, Malden, 373 pp

Sillitoe RH (2010) Porphyry copper systems. Econ Geol 105:3–41

Sinclair WD (2007) Porphyry deposits. In: Goodfellow WD (ed), Mineral deposits of Canada: a synthesis of major deposit-types, district metallogeny, the evolution of geological provinces, and exploration methods: Geological Association of Canada. Special Publication No. 5, pp. 223–243

Simmons S.F., Browne P.R., (2000) Hydrothermal minerals and precious metals in the Broadlands-Ohaaki geothermal system: implications for understanding low-sulfidation epithermal environments. Economic Geology (95) 971–999

Von Damm K.L., (1990) Seafloor hydrothermal activity: black smoker chemistry and chimney. Annual Reviews of Earth and Planetary sciences (18) 173–204

Chapter 5
Deposits Formed by Sedimentary and Surficial Processes

5.1 Introduction

We have chosen in this chapter to group several types of ore bodies that have the common characteristic of having formed at or very near the surface of the solid earth. One group of deposits results from sedimentation; i.e., the accumulation of detrital grains or chemical precipitates in rivers, lakes, coastal settings or in shallow to deep oceans. The other group comprises deposits that form in zones of weathering just below the surface, usually in humid tropical environments. Table 5.1 summarizes these processes and gives examples of each type of deposit.

In each case the minerals that are mined in these deposits are stable at low temperatures in the humid, usually oxidising environment at the surface of our planet. Three broad types of ore-forming processes can be identified.

1. A placer ore body is a concentration of eroded particles of valuable minerals in an alluvial or eluvial deposit of sand or gravel. The ore minerals in such deposits crystallize initially within the crust, in magmas or metamorphic rocks, or from hydrothermal fluids. In some cases their concentrations in these source rocks are higher than in normal crustal rocks, as in a gold vein or diamond-bearing kimberlite, but in other cases the minerals are accessory phases present in their usual concentrations. These minerals are released from their host rocks by uplift, weathering and erosion and are then concentrated in ore bodies by sedimentary processes. Placer gold deposits in rivers gravels or aeolian sands, and deposits of heavy Zr and Ti minerals in beach sands, are examples of this type of deposit.
2. Other ore minerals precipitate from lake water or seawater to form chemical sedimentary rocks. The metals or other valuable minerals in such deposits are soluble in surface waters but precipitate when they reach saturation levels or when the composition or physical conditions of the water changes. Examples include salt deposits that result from the evaporation of waters in lakes or shallow seas, and Fe- or Mn-rich sediments that result from mixing of waters with different compositions or redox states.

N. Arndt and C. Ganino, *Metals and Society: an Introduction to Economic Geology*, DOI 10.1007/978-3-642-22996-1_5, © Springer-Verlag Berlin Heidelberg 2012

Table 5.1 Deposits in surficial settings

Type	Subtype	Commodity	Process	Examples
Placer	Gold	Au	Accumulation of gold particles in river or beach gravels	California, Victoria, Klondike, Witwatersrand
	Zr-Ti minerals	Au	Accumulation of heavy minerals in beach sands	Western and Eastern Australia, South Africa, Florida
	Diamond	Au	Accumulation of diamond in near-shore gravels	South Africa, Namibia
Sedimentary	Ironstones (Algoman type)	Fe	Reworking and redeposition lateritic oolites	Lorraine, France; Minnesota, USA
	Iron-formation (Superior type)	Fe	Deposition of Mn chemical sediments	Hammersley, Australia; Brazil
	Mn sediment	Mn	Deposition of Fe-Si chemical sediments	Groote Island, Australia
	Evaporite	Salt (NaCl), potash (KCl)	Evaporation and depositions of chemical sediment	Various, Saskatchewan, Canada
	Evaporite	Nitrates	Evaporation and depositions of chemical sediment	Atacama, Chile
Calcrete	Uranium	U	Deposition of uranium in surficial deposits	Yeelerie, Australia
Laterite	Bauxite	Al	Lateritic soil on granite or clay-rich sediment	Jamaica; Cuba; Australia; Salindres, France
	Ni-laterite	Ni	Lateritic soil on ultramafic rock	New Caledonia
Supergene	Various	Various	Upgrading of metal concentrations in upper parts of ore deposits	Various

3. The third type of ore body is made up of minerals that are stable in zones of intense weathering at the surface of the Earth. Particularly in hot and humid conditions, many rock-forming minerals break down to form secondary phases that are soluble in surface waters. These compounds are removed by circulating groundwater, leaving only the insoluble minerals that become increasingly concentrated as the other minerals, which make up the bulk of the original rock, are leached out. Bauxite (the ore of Al) and Ni lateritic deposits form in this way. A similar type of process acts when other types of ore deposits are exposed at the surface; the soluble components are leached out leaving a residual

Table 5.2 Physical properties of minerals in placer deposits (From Garnett and Bassett (2005))

Mineral	Commodity	Density (g/cm^3)	Moh's hardness	Tenacity	Grindability[b]	Remarks
Diamond[a]	Diamond	3.5	10	Brittle	1	Very hard but only moderate density; brittle
Gold	Au	15–19	2.5–3.0	Malleable	11	Very dense, soft, malleable
Platinum	PGE	14–19	4.0–4.5	Malleable	10	Dense, moderately soft, malleable
Cassiterite	Sn	6.5–7.1	6–7	Brittle	6	
Rutile	Ti	4.2–4.3	6–6.5	Brittle	3	Heavy mineral in beach sands
Zircon	Zr	4.7	7.5	Brittle	2	Heavy mineral in beach sands
Monazite	Th, REE	5–5.3	5–5.5	Brittle	9	Heavy mineral in beach sands
Ilmenite	Ti	4.7	5.5–6.0	Brittle	8	Heavy mineral in beach sands
Garnet	Abrasive	3.5–4.3	6.5–7.5	Brittle	4	Industrial mineral

[a]Minerals arranged in order of their survivability in the fluvial environment
[b]Resistance to grinding in laboratory ball-mill tests

layer that may become enriched in valuable metals. This process is called supergene alteration or supergene enrichment.

5.2 Placer Deposits

A placer ore body is deposit of sand, gravel or soil containing eroded particles of valuable minerals. These minerals are able to survive and become concentrated in the surface environment because of their chemical and physical properties, as illustrated in Table 5.2. They are not necessarily thermodynamically stable but their rates of reaction are slow compared with the duration of erosion, sedimentary transport and deposits. An example is diamond, which, although thermodynamically unstable at low pressure (James Bond had it wrong, diamonds are not "forever"), is well able to survive long enough to be transported from their kimberlitic source to a site of deposition in offshore gravels. Other minerals in placer deposits are gold, which occurs in the native or metallic form, and oxides or silicates such as rutile, ilmenite, zircon, and monazite, which are the sources of Ti, Zr, Nb and other high-technology metals. Uraninite is stable only under reducing conditions such as those that existed in the atmosphere, oceans and rivers of the first part of Earth history, and placer deposits of these minerals formed only in the Archean and early Proterozoic.

Diamond, uraninite and the Zr-Ti oxides are moderately to extremely hard, which allows them to resist abrasion as the grains are released from their host

Table 5.3 Arguments for and against the hydrothermal and placer models for Witwatersrand gold deposits

Hydrothermal model	Modified placer model
Gold is late in the paragenetic sequence and is associated with zones of hydrothermal alteration	Coexistence of rounded gold nuggets and hydrothermal gold
Rounded pyrite grains and uraninite are of postdepositional hydrothermal origin	Pyrite morphology, crystallography and zonation patterns indicate detrital origin
Gold, and uranium, are closely associated with pyrobitumen that was remobilized during metamorphism	The association relates only to hydrothermally remobilized gold and uranium
Permeable conglomerate beds channelized the flow of hydrothermal fluids	Strong sedimentary control on gold distribution
Gold was deposited after sediment deposition	Re-Os ages of gold are older than the sedimentation
Gold was deposited during peak metamorphism	Metamorphism remobilized pre-existing detrital gold
Lack of suitable source of placer gold	Surrounding granites and greenstones were the gold source

Modified from Frimmel et al. (2005)

rocks and carried along in rivers or ocean currents. Gold, of course, is very soft but it is also highly malleable and ductile, which makes it durable in the sedimentary transportation (Table 5.3).

In addition to being resistant in the surface setting, most of the ore minerals in placer deposits are significantly denser than other minerals that are transported by sedimentary processes. It is this characteristic that allows these minerals to be sorted out from detrital minerals or rock fragments that make up most of the sediment load and become concentrated in ore bodies. Table 5.2 shows some examples: diamond is only slightly denser than detrital minerals like quartz or feldspar, and only its extreme hardness allows it to accumulate. Gold is the counter example, its density being six times greater than that of common rock-forming minerals. Because of the high density contrast, sorting by river currents allows even small particles of gold to separate efficiently from other detrital grains, to produce placer ore bodies that contain gold concentrations that are greater by a factor of several 1,000 than concentrations in normal continental crust. Sorting of moderately dense minerals like zircon and rutile is less efficient and placer deposits of these minerals are far less enriched, containing concentrations of Zr and Ti that are normally only 10–100 times greater than in normal crustal rocks.

5.2.1 Gold Placers

Placer gold deposits have produced two thirds of all the gold that has ever been mined. The easily won gold in the fluvial placers of California, Australia and elsewhere was mined out very rapidly in the gold rushes, usually over a period of

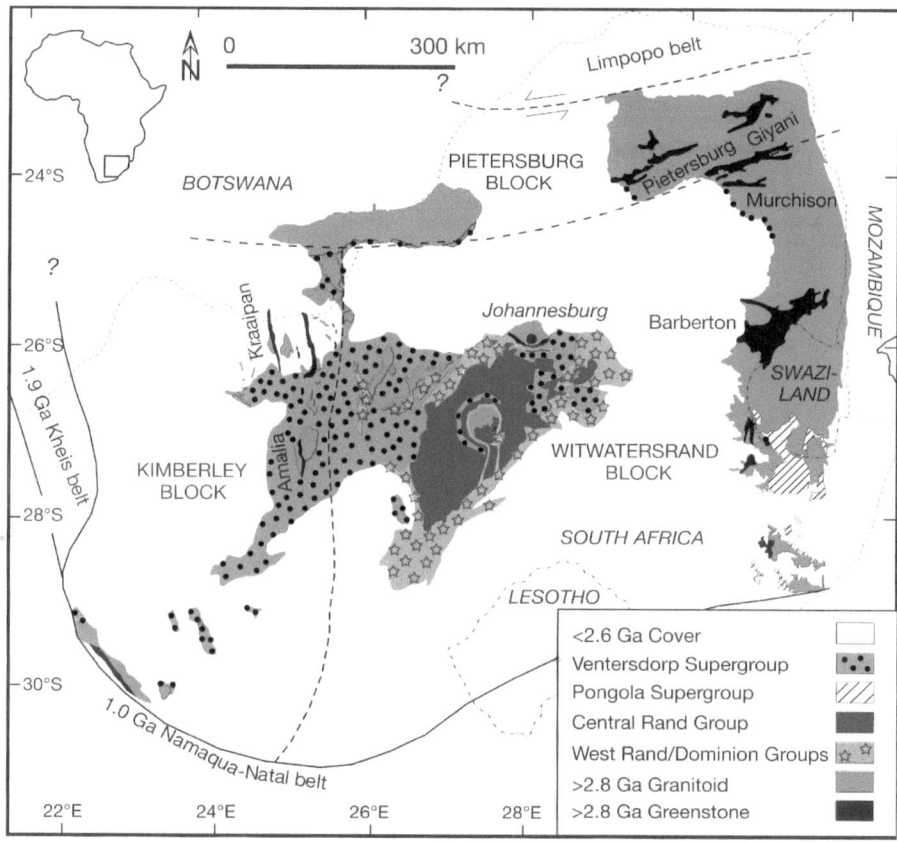

Fig. 5.1 Sketch map of the Witwatersrand basin, showing the central disposition of the sands and conglomerates of the Central Rand and West Rand groups, the flood basalts of overlying Venterdorp Supergroup and the surrounding granites and greenstones, which were the probable source of the gold (Modified from Schmitz et al. (2004) and Frimmel et al. (2005))

only a few years, and subsequent mining in each of these places became focused on "lode gold"; i.e. gold in veins in solid rock deposited from hydrothermal fluids, as described in Chap. 4. At present, gold production from placers is largely artisanal and makes up only a minor fraction of total gold production. On the other hand, production continues in the Witwatersrand ore bodies of South Africa (Fig. 5.1), a (hydrothermally reworked) conglomeratic paleoplacer deposit that is the largest gold deposit that we know of. The Witwatersrand ore bodies have produced nearly half the gold that has ever been mined. In 1970 they still accounted for half of global production but by 2007 the figure had fallen to 11% due to the exhaustion of shallow and rich ore bodies in the Witwatersrand and the development of other gold mines in many parts of the world. China is now the biggest producer.

Box 5.1 The Great Gold Rushes of the Nineteenth Century

Free gold, the term used to describe separate particles of native (elemental) gold in stream beds, is readily extracted by miners using simple methods such as the gold pan or sluice box (a simple wooden frame over which a slurry of gold-bearing gravel and sand is allowed to flow – ridges or mats on the floor of the box collect the particles of gold). Deposits of alluvial or placer gold were mined by many primitive societies, including the Romans who for two

(continued)

centuries mined the gold at Las Medulas in Spain after they conquered the region in 25 BC. Similar deposits triggered the gold rushes to the USA, Australia and the Yukon of Canada. Stories of miners making their fortune after finding enormous nuggets attracted hundreds of thousands of prospectors to the alluvial gold fields of California (1848–1852), Victoria (1851) and the Klondike (1898–1899). The "Welcome Stranger", the largest single lump of gold ever found, measured 61 cm by 31 cm and weighed almost 70 kg. This nugget was found accidently in a cart track in Victoria in 1869. Artisanal mining of placer gold continues to the present day in many parts of the world, notably in parts of South America (where the use of cyanide and mercury to extract the gold causes major pollution of river systems).

In these areas, which are located in or adjacent to young mountain ranges, the gold particles originally formed as lode gold deposits (Chap. 4) and were subsequently concentrated in the beds of streams and rivers that flowed rapidly in these hilly to mountainous terrains. Another types of placer deposit formed through alluvial processes in desert sands in the flat dry plains around Kalgoorlie and Coolgardie in Australia, fueling a gold rush to this region at the end of the nineteenth century. This gold came from enormous lode-gold deposits in the Archean greenstone belts of this region (Chap. 4), which have been mined continuously using conventional methods since that time. Similarly large and rich deposits in Canadian greenstone belts such as those around Timmins or Kirkland Lake yielded very little placer gold and triggered only minor gold rushes (the Porcupine gold rush in 1909–1911). This was in part due to the hostile winter climate but mainly because continental glaciers scoured this region in geologically recent time, removing any existing placer deposits and leaving a subdued topography in which modern rivers flow only sluggishly and concentrate little gold.

Young gold placers consist of accumulations of gold particles in Quaternary and Tertiary gravel, sand or soil, and their consolidated equivalents. Two broad types can be distinguished; alluvial deposits, in which the gold is transported by river or ocean currents and is separated from grains of other minerals by the action of these currents; and eluvial deposits, in which the gold remains more or less in place at the site of exposure of the primary deposit while the other minerals are removed. One type of eluvial deposit forms on hill slope immediately below the exposures of gold-bearing veins: gravity sliding or the action of wind or water removes the lighter components leaving the denser gold particles where they are. Another type forms on desert plains where winds or occasional floods remove the less-dense or soluble minerals.

In alluvial placers in river or stream beds, the flowage of water concentrates gold at locations where the velocity decreases markedly or where currents with contrasting flow rates or flow styles (laminar or turbulent) are juxtaposed. Examples include sand or gravel banks on the interior of meanders, below rapids and falls, downstream from boulders or other obstacles such as fractures, joints or ridges in streambeds. In all of these locations, the less dense or finer particles are transported away from the site of deposits by fast-flowing water leaving the denser or larger grains in regions of reduced water velocity.

Gold in placer deposits is almost entirely in the metallic form and occurs in a range of grain sizes, from minute particles or flakes to large nuggets. Gold in placers is very pure, normally 80–85% Au, the rest being mainly Ag. At locations where the original lode source of placer gold can be identified, the placer gold is normally purer than the lode gold. The refinement apparently takes place in part during oxidation when the lode gold is exposed at the surface, because silver is more soluble than gold under these conditions. However, fine layers of very pure gold (98–99%) commonly forms rims around particles and nuggets of gold in placer deposits, which leads to the hypothesis that the nuggets continue to grow within the placers themselves by some sort of dissolution-precipitation process.

Quartz pebble conglomerate or pyritic paleoplacers, of which the ore bodies in the Witwatersrand Basin in South Africa are the type example, were deposited in braided streams and alluvial fans during the Paleoproterozoic. The conglomerates consist of well-rounded pebbles of quartz, chert and locally pyrite (Fig. 5.2) in a matrix of quartz, mica, chlorite, pyrite and fuchsite. They contain native gold, pyrite and other sulfides, arsenides and sulfosalts, uraninite (UO_2), brannerite (U^{4+},Ca) $(Ti,Fe^{3+})_2O_6$, small nuggets of the platinum group elements and, significantly (for reasons explained below) pyrobitumen. The native gold is located in the matrix of the conglomerates and minor amounts of gold also occur in the pyrite and other sulfur minerals. Two morphological types of gold can be distinguished, well-rounded grains that resemble small nuggets in young placer deposits, and irregular aggregates and well-crystallized euhedral overgrowths that resemble gold of hydro-thermal origin.

Figure 5.1, a sketch map of the Witwatersrand basin, shows the central disposition of the sands and conglomerates of the Central Rand and West Rand groups that contain the auriferous conglomerate reefs. It also shows the overlying Ventersdorp flood basalts, which protected the reefs from erosion; and the surrounding granites and greenstones, which were the probable source of the gold. Sediments of the

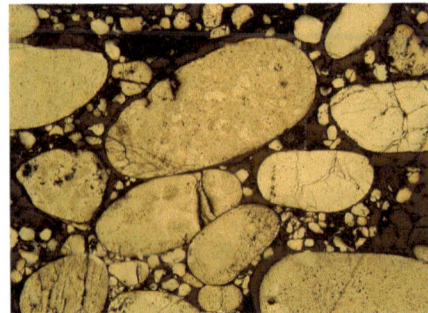

Fig. 5.2 Photograph of rounded grains of detrital pyrite in Witwatersrand conglomerates – evidence of reducing conditions during the formation of the deposits. The field of view is 8 mm (Photo A. Hofmann)

Central and West Rand groups were deposited between 2.8 and 3.0 Ga at the margin of an inland sea in a system of braided rivers that eroded material from the surrounding highlands. Fluctuations in sea level repeatedly changed the position of coastline, building up a ~5 km-thick sequence of deltaic sands, shales and conglomerates. The orebodies are located in lenses of pebbly arenite at several different stratigraphic levels in the Witwatersrand Supergroup. Other deposits or this type include those in the ca. 2.1 Ga Tarkwaian System of Ghana and at Jacobina, Bahia, Brazil.

Box 5.2 Exercise – Geological, Economic and Ethical Aspects of Gold Mining

Ore mined from the Witwatersrand deposits averages 10 ppm Au, 30 ppm Ag and 280 ppm U. In normal samples of ore in this deposit, as with most other modern gold deposits, the gold is invisible to the naked eye. 10 ppm is the equivalent to 10 g/t – a tonne of rock must be dug out and treated in order to extract a piece of gold the size of a sugar cube! Mining the gold produces enormous holes in the ground and vast waste dumps at the surface. If not properly maintained the dumps leak toxic metals into surrounding rivers and although they are usually replanted and transformed into grass- or tree-covered hills, the remediation process is long and not always effective.

On the other hand the total surface of land used in gold mining is not very large, the size of a large American or Australian wheat field. The farms are not environmentally benign – they have replaced the original forest of the area, they are ecologically sterile in that they support only one species, and they too generate pollution in the form of nitrates and pesticides.

The amount of wealth generated by the gold mine is many orders of magnitude greater than that produced by the farm. For example, the mine site, tailings dumps and all other infrastructure of Telfer, a moderately sized gold mine in Western Australia, cover about 30 km^2. The mine produces about 1,000,000 oz of gold/year, which, at \$1,200/oz, is worth 1.2 billion dollars. Several big wheat fields cover about the same area (30 km^2) and produce about 800 t of wheat/km^2/year, a total of 24,000 t. The wheat price is about \$100/t giving a production worth 2.4 million dollars. In other words, the wealth produced by the gold mine is 500 times greater that that from the wheat field.

The Telfer mine employs about 1,000 workers, the large wheat fields about 50.

Some 200 miners lose their lives each year in South African gold mines and the health of thousands of others is permanently damaged by silicosis and other diseases contracted during the life down the mines. Artisanal mining pollutes streams and rivers through South America.

Some of the gold produced in the mines is indeed used in industry (the computers we are using to write this book each contain about half a gram), but most newly mined gold is used in jewelry, particularly in India where to metal

(continued)

is highly prized, or stays locked up in the vaults of central banks. Gold is also a "crisis metal", a refuge that investors seek in times of economic turbulence. During 2008–2009, as the price of ferrous and base metals plunged, the price of gold approached record heights. The high values have been maintained during the following years of global uncertainty.

How can we balance the positive and negative aspects of gold mining? Discuss the geological, economic and ethical aspects of this activity.

Although we have chosen to describe the deposits of the Witwatersrand basin in this chapter of deposits in sedimentary rocks, the origin of these deposits is controversial. There are two competing hypotheses for the origin of gold, the *placer model* and the *hydrothermal model*. Arguments for and against each are summarized in Table 5.1.

According to the placer model, detrital grains of gold and were transported into the basin and deposited in the matrix of the conglomerates. Figure 5.2 shows rounded grains of detrital pyrite in the Witwatersrand conglomerate. Under the modern oxygen-rich atmosphere these grains would be replaced by iron oxides and could not survive in fluvial conditions. Their presence is taken as evidence for more reducing conditions in the later Archean.

According to the hydrothermal model, hot $H_2O–CO_2$ fluids, perhaps derived from dehydration of metavolcanic rocks beneath the basin, flowed along the permeable conglomeritic horizons and deposited gold and other minerals in the pore space of these sediments. Proponents of both schools agree that hydrothermal fluids did indeed flow through the conglomerates on one or more occasions: the hydrothermal school believes that most or all of the gold was introduced in these fluids, whereas the placer school believes that these fluids caused relatively minor recrystallization and redistribution of gold of detrital origin. Because of this remobilization, the latter model is commonly referred to as the "modified placer model", as in Table 5.1. Both groups have developed numerous lines of evidence and yet, despite many decades of research and argument, the issue is still not resolved. Just as with another major South African ore type – the chromite and PGE deposits of the Bushveld (Chap. 3) – neither the fame of the deposits, nor their economic importance, nor long years of research, have been able to resolve outstanding questions surrounding the origin of the ores!

5.2.2 Beach Sands

Most of the world's supplies of titanium and zirconium come from concentrations of heavy minerals in beach sands. These elements are known "high-technology" or "space-age" metals because their high strength-to-weight ratio makes them very suitable for the construction of aeroplanes and spacecraft, not to mention golf clubs, the casings of Macintosh computers and the masts of racing yachts. The major use of titanium is not as a metal, however, but as an oxide: TiO_2 is the pigment that imparts a brilliant white colour to a wide variety of paints, papers, plastics and other materials. The rare-earth elements, which are finding increasing uses in superconductors, ceramics, batteries, magnets, phosphors in TV screens, and catalysers in petroleum refineries, are present in monazite, a minor component in beach sands.

Heavy mineral deposits in relatively young beach sands are the only significant source of zirconium, in addition to being a major source of titanium. The latter metal is also mined in magmatic deposits of ilmenite ($FeO.TiO_2$) in anorthosites in Canada and Norway and as anatase (TiO_2) in residual deposits overlying alkaline intrusions in Brazil.

The principal ore minerals in heavy-mineral beach sands are ilmenite ($FeO. TiO_2$), leucoxene (approximately TiO_2, an alteration product of ilmenite), rutile (TiO_2), zircon ($ZrSiO_4$) and monazite (a phosphate of Th and the rare earth elements). The concentrations of ore minerals are highly variable. The richest deposits contain up to 50% of ilmenite, 5–20% of both rutile and zircon and 1–3% of monazite; more normal ores contain only a few percent of combined heavy minerals. However, since the deposits are located at the surface in the form of unconsolidated sands, mining costs are much lower than those of normal deposits in which the hard solid rock must first be extracted from deep mines then crushed before the ore minerals can be extracted.

All the heavy minerals in beach-sand deposits occur as accessory phases that are present in small quantities ($<1\%$) in normal felsic magmatic and metamorphic rocks. Grains of these minerals are released when the host rock is broken down by subaerial weathering and these grains, which are stable in the low-temperature fluvial environment, are then transported by streams and rivers to the shorelines of lakes or oceans where the action of waves, wind, tides or long-shore currents winnows out the lighter quartz and feldspar, leaving the sands enriched in the dense Ti and Zr minerals. For the process to be effective, a large reservoir of the source rock must be subjected to long periods of chemical weathering. This is the situation in old Precambrian cratons which are rich in granitic or gneissic rocks that contain relatively high concentrations of Zr and Ti minerals, and which are eroded down to a peneplane that is subject to protracted and intense weathering, particularly when located in equatorial regions. When similar rocks are exposed in younger mountain ranges in temporal or cold climates, mechanical weathering dominates and the heavy minerals are not so effectively separated from other components. For these reasons, the largest deposits are located around old stable continents in equatorial regions, namely along the coastlines of Australia, South Africa and India. These countries are therefore the major producers of Zr and Ti. Smaller or less rich deposits are located in southern USA, West Africa, Malaysia and China.

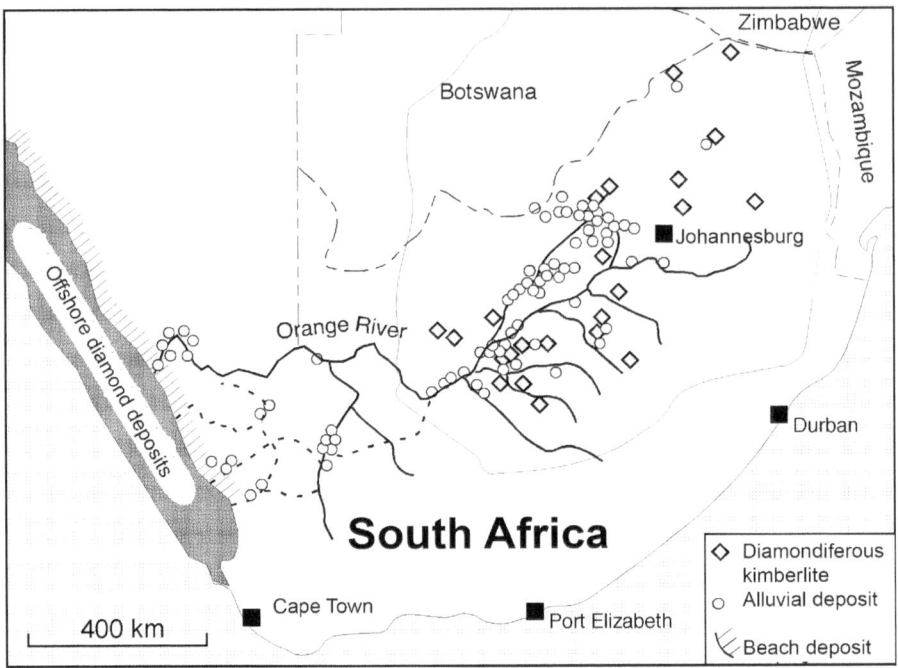

Fig. 5.3 Location of three different types of diamond deposit in southern Africa: primary kimberlitic diamond; alluvial deposits in present and past river beds, and alluvial deposits in offshore gravels (From Lynn et al. (1998))

5.2.3   Alluvial Diamonds

The first diamonds discovered in South Africa were in gravels of the Orange River and its tributaries. Tracing these rivers back to their sources led first to the discovery of the primary diamond sources in kimberlites around the town of Kimberley in the centre of South Africa, and then to enormous beach placers at the western coast of the continent in South Africa and Namibia (Fig. 5.3).

Placer deposits are the source of about 34% of global diamond production. By far the largest fields are those along the western coast of southern Africa, from Cape Town to the Congo, but placer diamonds have been mined in India throughout history and also in large quantities in Russia and Brazil. Placer deposits normally contain a high proportion of high-quality gem diamond (up to 97%) because large flawless single crystals survive the rigours of sediment transport and deposition better than smaller flawed crystals.

Alluvial diamonds in the Orange River and tributaries (Fig. 5.4) are mined at scales from artisanal to medium-sized commercial operations. Recovery of diamonds in offshore gravels, which are present at water depths up to several 100 m, requires the heavy equipment – a diverse flotilla of barges connected to large tubes that suck the diamonds from the sea floor (Fig. 5.5).

Fig. 5.4 Photos of alluvial diamonds (Source: www. diamondfields.com)

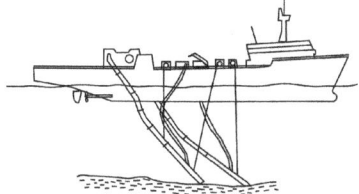

Air lift system is one of the method deployed for marine diamond mining. A vacuum is created using compressed air in order to suction the sediment to an onboard separation plant.

Fig. 5.5 Techniques used to extract diamonds from offshore deposits

5.2.4 Other Placers: Tin, Platinum, Thorium-Uranium

The primary source of tin is cassiterite (SnO_2) a magmatic mineral that crystallizes in an unusual type of granite. The so-called "tin granites" have a rather restricted distribution, both in space and geological time; they are very rare in the Precambrian and become more abundant in later epochs. The best known examples are on the Malayan Peninsula, an endowment that has made Malaysia the world's greatest producer of the this mineral. About half the deposits are in the granites themselves and the other half are in placers in rivers, beach sands and offshore deposits.

Much the same story applies to the platinum-group elements. The primary source is in ultramafic rocks such as those of the Bushveld Complex (Chap. 3) but in certain regions, nuggets of PGE alloys released by weathering and erosion of ultramafic rocks have produced viable placer deposits. In some cases the sources are ore deposits in their own right, as in the ultramafic complexes of Siberia; in other cases ophiolites with no commercial ore concentrations have yielded the ore minerals, as in New Caledonia.

Thorium and uranium are hosted in uraninite in the Witwatersrand ore bodies which, as discussed above, may be of alluvial origin, and in monazite in beach sands. The beaches flanking some parts of India contain high concentrations of monazite and the country is currently developing a new type of nuclear reactor that uses Th as a fuel.

5.3 Sedimentary Fe Deposits

5.3.1 Introduction

Although the global production of some commodities may have reached a maximum and may now be passing into decline, petroleum being the key example, this is far from the case for iron. World consumption of iron ore was about 1,400 t in 2006 and even allowing for major increase due to growing demand in China and other developing countries, the annual figure is unlikely to exceed 2,000 t/year (the proportion of recycled iron, currently 8–10%, is likely to increase). Global *reserves* of iron ore (deposits whose existence has been proven by drilling) are estimated at an enormous 800,000 t of ore, and global *resources* (ores whose existence can be inferred) are more than double this amount. Dividing reserves or resources by the annual rate of consumption gives us the period of time before iron ores will be exhausted – 400 or about 1,000 years, depending on which figure is used. How much iron will mankind need at the start of the next millennium?

On the other hand, as with most other commodities, iron ore deposits are not distributed equally across the globe. China, the world's biggest producer, mined over 260 t of iron ore from its vast domestic resources in 2007; but it is also the biggest consumer, thanks to the growing demand for the steel needed to build in infrastructure that is required for a rapidly industrialising society. Currently China imports about 500 t of ore each year, mainly from Australia and Brazil. India is going through a phase of reindustrialisation similar to that in China, but many geologists and politicians believe that reserves of iron ore in that country are relatively limited. The question of how to manage mineral resources is the topic of the following exercise.

Box 5.3 Exportation Versus Conservation of Mineral Resources. Policies and Consequences in Australia and India

In the summer of 2009 (when we wrote the first version of this book) the Indian government was giving serious thought to banning the exportation of iron ore. The argument was that all known deposits in the country were relatively modest and should be reserved for domestic consumption by future generations.

In Australia in the 1950s, it was believed that reserves of iron ore in the country were very small; sufficient only for domestic use. Exportation of iron ore was forbidden, mineral companies had no interest in exploring for new deposits and none were found; the situation persisted for a decade.

On 16 November 1952 Lang Hancock, a prospector and cattle rancher, was flying over the Hamersley Range in rugged outback of the Pilbara in the northwestern part of the country. Bad weather forced him to fly at low altitude and as he did he wondered if the bands of bright red rock visible in nearby valley walls might be iron ore. A year later he returned and found that they

(continued)

were the exposed parts of a huge iron deposit, a discovery that led to the establishment of the iron ore mines in the Pilbara. Now, some 50 years later, Australia is known to have huge reserves of iron and is the world's major exporter of iron ore.

Question: how should governments manage the development of mineral resources? Should they guard known resources for future generations by limiting the rate of exploitation and banning exports, or should they free the market and encourage exploration in the hope of finding new deposits?

5.3.2 Types and Characteristics of Iron Deposits

A combination of two processes produces the richest iron ores. Fe minerals first precipitate from seawater to form chemical sediment; then chemical weathering upgrades the Fe content when the deposit is exposed at the surface. (The latter process, called supergene enrichment, is described later in this chapter).

There are several different types of iron deposit, as listed in Table 5.4. The type known as "ironstone", "minette" or "Lorraine type", has been exploited throughout Europe since the start of the iron age around 800 BC. Mining of the iron ores of the

Table 5.4 Types of sedimentary iron deposits

1. *Banded Iron Formations*

   Types

   Algoman

   Deposits of Archean age, in greenstone belts associated with volcanic rocks; Fe or probable volcanic exhalative origin

   Superior

   Early Proterozoic age, on stable continental platforms; deposition of Fe from ocean water at the time of increasing oxygen content of atmosphere and oceans

   Rapitan

   Neoproterozoic deposits associated with glaciogenic sediments

   Facies

   Oxide

   The most important type – the Fe mineral is hematite or magnetite.

   Unmodified sedimentary rocks contain up to 30–35% Fe

   Carbonate

   Alternating bands of chert and siderite ($FeCO_3$)

   Silicate

   The Fe minerals are silicates such as greenalite (a serpentine or clay with the composition $(Fe^{2+},Fe^{3+})_2$-$3Si_2O_5OH_4$), chamosite (phyllosilicate (Fe;Mg;Fe) Al(Si Al)O (OH;O)), glauconite ((K;Na)(Fe;Al;Mg) (Si;Al) O (OH))

   Sulfide

   Alternating layers of pyrite and shale rich in organic matter

2. *Ironstones (Minettes or Lorraine-type)*

3. *Bog irons*

Lorraine basin and nearby deposits of coal (needed to smelt the iron ore) in northern France, Belgium and Britain fuelled the industrial revolution. Germany's designs on the Lorraine iron resources were one of the causes of the First World War.

Ironstones from the Lorraine region were known as "minette" or "small ore" because of their low iron content. The grade was typically around 30%, about 20–35% less than in "fer fort", the richer ore mined in other countries. In addition, the iron is present in silicate minerals that are difficult to refine, and are accompanied by phosphorous and other elements that further complicates the extraction processes. With the discovery of the vast ore deposits in Brazil and Australia, mining of European ores has became uncompetitive and one after another the mines in the Lorraine basin have closed, the last in the late 1990s.

Ironstones occur as lenticular beds commonly associated with organic-rich black shales. They contain little or no chert and the iron is present in minerals such as hematite (Fe_2O_3) goethite (FeO.OH) associated with carbonate (siderite, $FeCO_3$) and silicates such as greenalite (a serpentine or clay with the composition (Fe^{2+},Fe^{3+})$_{2-3}Si_2O_5OH_4$), chamosite (a phyllosilicate with the composition (Fe^{2+},Mg,Fe^{3+}) Al(Si Al)O (OH,O)), and glauconite (K,Na)(Fe,Al,Mg)(Si,Al)(O,OH). The iron oxides and chamosite are often oolitic. In addition to the European deposits, ironstones are known and mined in central USA where they are known as Clinton-type ores. They were deposited two well-defined age brackets, in the Ordovician-Silurian and again in the Jurassic.

Ironstones are thought to form when iron on the continents is subject to deep weathering in a warm humid climate, conditions that lead to the development of lateritic soils. The initial iron enrichment and the growth of small round structures, pisolites, took place in highly oxidized surface layer in response to low-temperature and chemical and biogenic processes. The lateritic soils were then transported into shallow waters in deltas or estuaries where current and wave action sorted and concentrated the iron minerals.

Banded Iron-Formations are the dominant type of iron deposit. The term is used for bedded chemical sediments comprising alternating layers of iron minerals, usually oxides or hydroxides, and fine-grained quartz or chert (Fig. 5.6). The banding is manifested at different scales: centimeter-thick beds of iron minerals and chert are internally divided into millimeter or sub-millimeter lamellae of the same minerals. Detrital components such as clays, or grains of quartz and feldspar, are usually rare. In the major iron formations of the Hamersley Basin, the continuity of the bedding is remarkable: a single 2.5-cm-thick band has been traced over an area of 50,000 km^2 and varves at a microscopic scale are continuous for 300 km.

Banded iron-formations (BIF) were deposited at three different time periods, all in the Precambrian. The oldest is 3.5–2.7 Ga; the second and by far the most important from 2.5 to 2.0 Ga; and the third, far less significant, from 1,000 to 500 Ma. The tectonic setting of deposition and the types of associated rocks is different in each case, and this has given rise the following names for each type: Algoman, Superior and Rapitan, respectively.

Algoman-type deposits are relatively small and are found in Archean greenstone belts in association with volcanic rocks. Superior-type deposits are named after Lake Superior in between Canada and USA where they were first mined and

Fig. 5.6   Photos of banded iron formations. (**a**) outcrop of BIF in the Mt Tom Price mine of the Pilbara, Western Australia; (**b**) General view of the mine (Photos from K. Konhauser)

studied. The North American deposits have since been eclipsed by the much larger deposits in the Hamersley Basin of Western Australia, the Transvaal Basin of Brazil, the "Quadrilatero Ferrifero" in Brazil, Krivoy Rog in the Ukraine and the Singhbhum region of India. Finally, Rapitan-type deposits, named after the type locality in the McKenzie Mountains of northwest Canada, are a relatively minor type associated with Neoproterozoic glacial deposits.

A parallel classification of BIF deposits is based on the mineralogy of the iron phases. Oxides (hematite or magnetite) are the main phase in most banded iron-formations, but in others carbonate, silicates, or sulfide predominate. The different types of iron minerals, whose compositions are given in Table 5.4, are stable under contrasting conditions of Eh and pH, from relatively oxidising to highly reducing, which suggests that they were deposited in different environments. As seen in Fig. 5.7, the oxides are stable in acid solutions, hematite at high Eh and magnetite at low Eh; and pyrite and siderite are stable in near-neutral solutions at low Eh (but these fields expand at high activities of carbonate or sulfur). In alkali solutions, ferrous iron is soluble over a wide range of Eh but the field is broader under reducing conditions. It can therefore be seen that an increase in Eh, oxidation, stabilizes hematite relative to Fe^{2+}, and this process is the key to formation of most BIF deposits.

The restriction of this type of deposit to the period 2.4–2.0 Ga, and the enormous quantity of Fe tied up in these deposits, is commonly taken as evidence that the composition of the Earth's oceans and atmosphere changed markedly at this time. The Archean, pre-2.7 Ga, atmosphere was a reducing, oxygen-free mixture of nitrogen, carbon dioxide and methane, rather like the atmosphere of Jupiter; the

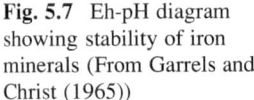

Fig. 5.7 Eh-pH diagram showing stability of iron minerals (From Garrels and Christ (1965))

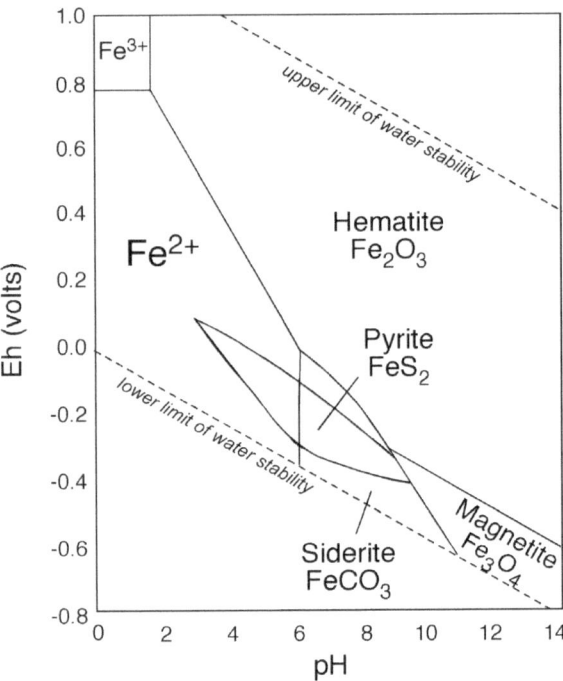

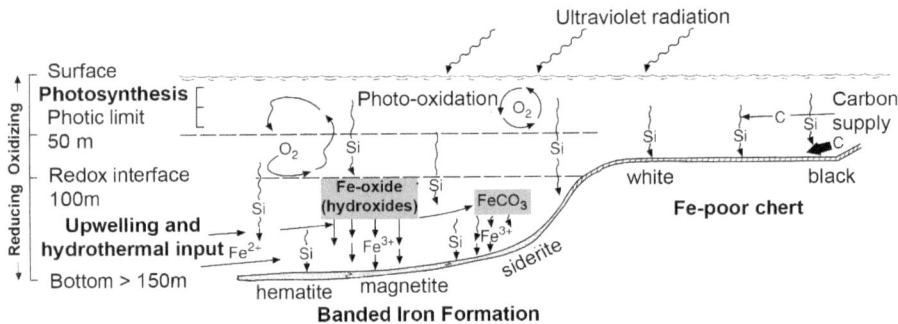

Fig. 5.8 Model for the formation of iron formations from Klein and Beukes (1993)

oceans were hotter, acidic and they contained abundant dissolved Si and Fe. In the early Proterozoic, the oxygen content of the atmosphere and the oceans built up, because or, or in parallel with, the appearance of abundant oxygenic cyanobacteria, and this change led to the precipitation of Fe and Si oxides. Figure 5.8 shows in more detail how the process might have worked during chemical or biochemically aided precipitation in shallow water on continental shelves or intracratonic, often evaporitic basins. In this model we assume that the iron is derived from hydrothermal input implying an exhalative volcanic source, but other models propose a source via weathering of continental or oceanic crust. Whichever the case, the iron minerals and silica are thought to precipitate when reduced Fe-rich deep seawater mixes with oxygenated shallow seawater. The different types of iron minerals are ascribed to different depths and conditions of deposition. Fluctuations in the flux of incoming fluid explain the centimeter-scale banding. As for the sub-millimeter scale laminations, it has been suggested that these reflect a cycle of heating, evaporation and oxygenation during the day and subdued activity during the night. In other words, the thickness of the lamellae could perhaps reflect the length of a day, more than two billion years ago!

Algoman-type iron formations probably result from more local mixing of reduced and oxidized fluids in restricted basins adjacent to Archean continents. An association with carbonate and sulfide facies suggests that the source of the iron were exhalative hydrothermal fluids emanating from the oceanic crust. *Rapitan-type* deposits are thought to form following "Snowball Earth" periods. This term is applied to periods during the Proterozoic when the entire planet was covered blanketed by ice sheets that covered both continents and oceans. Seawater became reducing, rather as in Archean times and was able to dissolve ferrous iron. During interglacial periods the reduced seawater mixed with oxidized surface waters, leading to the deposition of iron formations.

Primary iron formations contain 20–30% Fe whereas the ores mined in most countries contain 55–65% Fe. Enrichment processes that act on the iron formations after they are accreted to continents and as they are exposed at or near the surface explain the difference. Exposure in hot, humid climates to circulating groundwater leaches silica from the rock and replaces it by iron oxides. In the Hamersley Province in the Pilbara district of Western Australia, for example, three main types of enrichment process are recognized: supergene enrichment; dissolution of iron from the BIF and its redeposition as iron oxides along ancient, mainly Tertiary river channels; erosion, transport of fragments of iron deposits and redeposition of this detrital material in secondary deposits.

5.3.3 Other Sedimentary Deposits: Mn, Phosphate, Nitrates, Salt

Bedded deposits of Mn form in a manner very similar to iron formations. The ore minerals, pyrolusite, an oxide (MnO_2) or rhodochrosite, a carbonate ($MnCO_3$), precipitate from seawater as bedded sedimentary rocks. Controls on the solubility of Mn are like those of Fe: the metal is soluble in acid reducing fluids and its precipitation is caused by an increase in alkalinity or by oxidation. Manganese deposits are commonly associated with iron deposits and in several cases it is believed that initial precipitation of Fe leaves the water enriched in Mn that subsequently precipitates as the degree of oxidation increases. This process can be observed in the Black Sea where highly reduced (euxinic) deeper waters precipitate pyrite-rich muds as more oxidized surface waters precipitate Mn oxides.

Manganese deposits occur in rocks of all ages. The largest deposits are the Proterozoic ore bodies of the Kalahari in South Africa, the Cretaceous ores of

Groote Eylandt in Australia and the rhodochrosite ores of the Molango district of Mexico.

Phosphorites, which are mined to be used as fertilisers, form on shallow continent shelves either through direct precipitation from seawater or by diagenetic replacement of limestones. Biological processes are important in controlling the buildup of dissolved phosphorus in seawater and its subsequent precipitation.

The evaporation of brines in estuaries or lakes produce chemical precipitates that are mined for normal salt (NaCl), sylvite (KCl, a source of potassium in fertilisers), gypsum ($CaSO_4.2H_2O$, used in construction) and anhydrite (used in cement).

Sodium and potassium nitrates find a remarkable list of uses: as a fertilizer, a solid rocket propellant, a rust inhibitor and in gunpowder; in food preservation, in glass and pottery enamels; in the manufacture of cigarettes to maintain an even burn of the tobacco, and in toothpastes for sensitive teeth. The world's largest natural deposits of sodium nitrate are in the Atacama Desert of Chile, one of the driest regions of the world. The mining of these generated great wealth for Chile for over a century, until the 1940s when the German chemist Fritz Haber developed a process to produce ammonia from atmospheric nitrogen.

5.4 Laterites

Laterites are soils that develop through prolonged and intensive rock weathering in hot humid climates. Most laterites are relatively rich in iron and have no economic value, but when they form on granites or clay-rich shales, they form bauxite, the ore of aluminium, and when they develop on ultramafic rocks, they may form lateritic nickel deposits.

5.4.1 Bauxite

Table 5.5 is a list of the world's largest producers of aluminium metal. Included in the list are the large industrial countries like China and the USA and several countries with large reserves of bauxite, like Australia, India and Brazil. But also included in the list are small countries with no bauxite at all. Iceland, for example, is a small, sparsely populated volcanic island in the middle of the frigid North Atlantic – what is it doing in the list?

The reason is simple: to extract aluminium metal from the oxides and hydroxides that are the ore minerals in bauxite requires a lot of electricity, a commodity that Iceland has in excess. So bauxite or its refined product alumina (Al_2O_3) is shipped halfway around the world from Australia, refined in Iceland into aluminium metal which is then shipped to markets in Europe and America. Does this make any economic sense? And what is the impact on the environment of these "bauxite miles"?

Table 5.5 List of aluminium-producing countries (2010)

Rank	Country	Production (t)
1	People's Republic of China[a]	16,800
2	Russia[a]	3,850
3	Canada	2,920
4	Australia[a]	1,950
5	USA	1,720
6	Brazil[a]	1,550
7	India[a]	1,400
8	United Arab Emirates	1,400
9	Bahrain	870
10	Norway	800
11	South Africa	800
12	Iceland	780

[a]Major bauxite-producing countries
Source: http://minerals.usgs.gov/minerals/

In Iceland the cost of electricity is about 2 US cents/kWH (compared with up to 7 c/kWH in Germany). In Australia electricity is also rather cheap 3–4 c/kWH, not much higher than in Iceland, but the direct costs of shipping the alumina are not enormous and on balance the trade is viable. But the real advantage is seen when we consider the environmental issues. In Iceland the electricity is produced in hydro or geothermal energy plants that release almost no CO_2 into the atmosphere whereas in Australia 76% of electricity is produced from coal, the worst greenhouse-gas-emitter of all. Just as with the argument about whether it is wiser for Europeans to eat tomatoes grown in heated greenhouses in Holland or shipped from the other side of the Mediterranean, the ecological consequences are not immediately obvious.

Box 5.4 Carbon Footprint of Aluminum Production in Iceland from Bauxite Shipped from Australia

Transport by ship produces about 25 g of CO_2/t/km. This estimate does not take into account the carbon footprint of the construction, maintenance and dismantling of ships, but gives a rough idea of the carbon emissions of the shipment. Electricity from coal produces about 1,000 kg of CO_2/kWh, while electricity from geothermal or hydropower produces virtually none. Four tons of bauxite and 300 kWh are needed to produce 1 t of aluminum. The round-trip by sea from Australia to Iceland is about 40,000 km.

Use these figures to calculate if it is more advantageous, in terms of carbon emissions, to produce aluminum in Australia (using electricity produced from coal), or to export and refine bauxite in Iceland (using electricity produced without emitting CO_2). Watch out for units in your calculations! Your results will not take into account the lower cost of electricity in Iceland and the shorter shipping distances of the refined metal to consumers in European and American.

(continued)

Solution: When producing aluminum in Australia the transport costs are negligible but to refine 1 t of metal from 4 t of bauxite consumes 300 kWh of coal-produced electricity, which emits 300 t of CO_2. Transportation to Iceland emits only $40,000 \times 25 \times 4$ g or 4 t of CO_2 and emissions from electricity production are negligible. From this back-of-the-envelope calculation, it is clear that it is ecologically sound to ship ore half-way across the world to benefit from the low cost – both financial and in terms of CO_2 emission – of electricity in Iceland.

Just as for iron, there is little chance we will ever exhaust our resources of aluminium. The reserves of bauxite are enormous, about 25 billion tons, enough to last more than 300 years even allowing for greatly increased consumption. And even if bauxite were ever totally mined out, even larger amounts of Al are present in clay and feldspar, whose abundance is immeasurable. Even now some aluminium is produced in Russia by mining feldspar in alkaline intrusions.

The world's bauxite deposits are found mainly in equatorial regions with tropical climates; Guinea, Australia, Brazil and Jamaica. The biggest and richest deposits form on continental peneplains that are subject to long periods of alternating wet and dry seasons, conditions that cause the water table to oscillate. As its level of groundwater rises and falls, the more soluble components are leached out and what is left is a lateritic soil that retains only the most insoluble components. In parts of Africa, South and Central America and Australia, lateritization on flat peneplained surfaces has persisted for up to 100 million years resulting in a thick deeply weathered lateritic blanket whose thickness may reach 150 m.

The majority of major rock-forming elements – Si, Fe, Mg, Ca, Na and K – are moderately to highly soluble in the near-neutral (pH 5–9) waters in lateritic weathering horizons. The oxides of these elements make up about 80% of feldspathic rocks such as granite, gneiss or shale and if they were totally removed, the concentration of Al_2O_3 increases fourfold, from about 15% (the level in the source rock) to close to 60%, the level in rich Al ore. The best bauxites are mixtures of Al hydroxides – gibbsite ($Al(OH)_3$), and two polymorphs with the composition AlO (OH), boehmite and diaspore. In practice, the removal of the more soluble components is not complete; some of them are retained in partially altered rock and others accumulate, often in secondary minerals, at various levels in the lateritic soil profile.

In Fig. 5.9 we show a profile through two lateritic ore deposits, one containing bauxite and the other a Ni laterite. The saprolith zone at the base of the profile consists of highly weathered rock that preserves much of its original texture and structure. Feldspar and the ferromagnesian minerals are destroyed and the soluble components are partially removed while Si and Al are retained in clays and some Fe is retained in goethite and hematite. In the upper part of the saprolith and in the pedolith all but the most resistant primary minerals are destroyed and the rock in

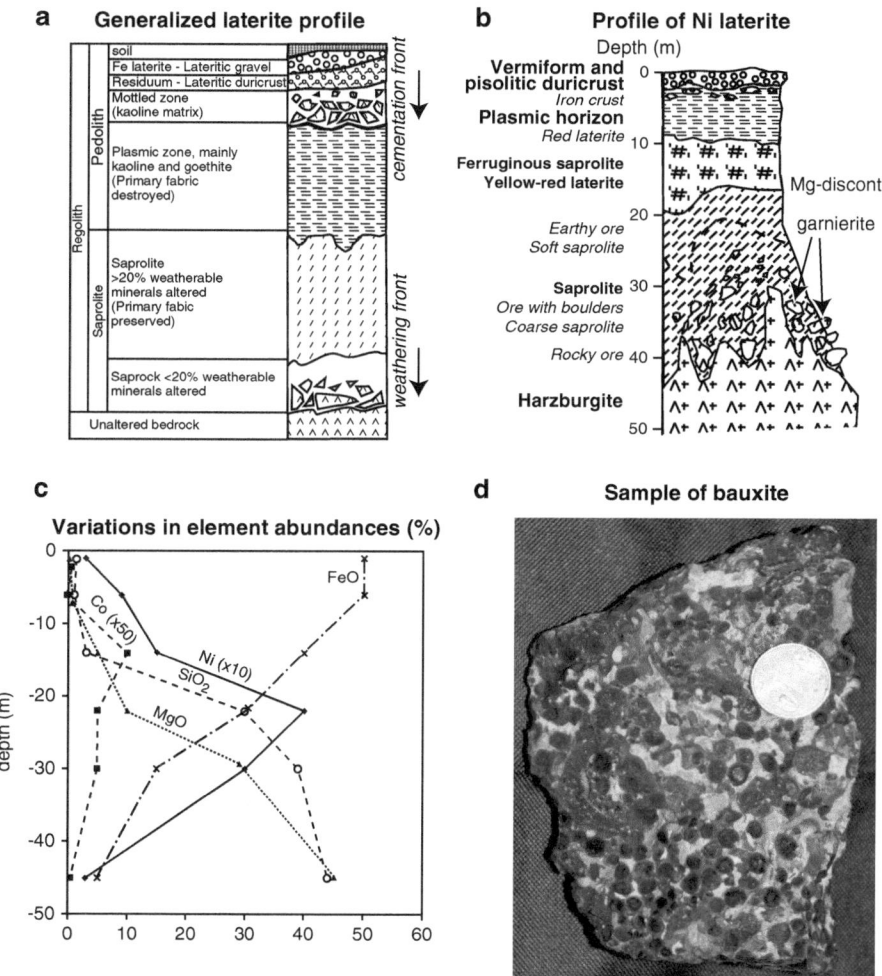

Fig. 5.9 Profiles through lateritic profiles (**a**) bauxite (From Butt et al. 2000); (**b**) nickel laterite (From Freyssinet et al. 2005); (**c**) chemical profile through the laterite, plotted from data of Freyssinet et al. (2005); photo of pisolitic bauxite

composed entirely of clays, Al and Fe oxides and hydroxides, and minor amounts of residual quartz. The original texture is completely lost and the rock has a pisolitic or nodular structure that results from repeated episodes of dissolution and accretion. The uppermost pedolith consists of unconsolidated or cemented Fe-oxide-rich gravels and duricrust.

Most laterites in equatorial regions have the iron contents which makes them unsuitable for the recovery of aluminium. The purest bauxites form through a combination of processes; (1) the presence of Al-rich (and Fe-poor) parent rocks such as alkali granite, syenite, tuff or clay-rich sediment and their metamorphosed equivalents; (2) an appropriate balance of temperature and rainfall (high temperatures favour Fe-rich laterites); and (3) a pronounced alternation of wet and dry seasons.

5.4.2   Ni Laterites

If it is accepted that an island half way across the world is part of the French nation, then the biggest mines in France are in New Caledonia. The nickel deposits of the island were found in 1864 by Jules Garnier who gave his name to the main ore mineral, a mixture of pale green or apple-green phyllosilicates with the approximate composition mineral, $(Ni, Mg)_3Si_2O_5(OH)_4$. The deposits have been mined almost continuously since 1875 and have seen the development of many innovative procedures for the extraction of Ni from lateritic ores. During the early years very rich garnierite ores containing up to 15% Ni and averaging 2.5% Ni were exploited but these are now largely exhausted and mining has turned the less rich goethitic ores (1.3–1.6% Ni). The long and involved history of the development of a new deposit in the Goro mine, from its conception in 1993 to planned production in 2013, is given at http://www.valeinco.nc/pages-eng/propos/history.htm. The New Caledonian deposits are thought to contain about a third of the world's supply of Ni; and when other major lateritic deposits in Indonesian, Cuba and Australia are included, the figure reaches 50–60%.

Just as with bauxite, the extraction of Ni metal from the silicates and hydroxides that constitute lateritic nickel ore requires abundant energy. The lateritic ores contain abundant bound water, which must be driven off, and have high magnesium contents which means that conventional smelters must run at very high temperatures. A wide variety of alternative extraction processes have been developed New Caledonia and Cuba and these are being applied with variable success in the extraction of ore in other countries. Because of the high energy consumption, the viability of a nickel laterite project depends crucially on the price of energy, which is dictated by the oil price; when it is low, lateritic deposits gain a competitive edge over magmatic deposits. Table 5.6 lists the advantages and disadvantages of the two types of Ni deposit.

Nickel laterites develop when ultramafic rocks are exposed to protracted weathering in hot humid climates. In most of areas where deposits are known – New Caledonia, Indonesia, Cuba, etc. – the ultramafic rocks come from the lower, mantle portions of ophiolites. A notable exception is the deposits in Western Australia which have formed on intrusions of komatiitic lineage, the same rocks that host magmatic ore deposits.

The laterization process is directly comparable to that of bauxite formation and the parallel is brought out in the comparison of the two profiles in Fig. 5.9. Nickel laterites develop when either the primary minerals of the ultramafic rocks, olivine or pyroxene, or secondary serpentine that replaces these minerals, are subjected to weathering under conditions similar t those that produce bauxite. Elements such as Si, Mg, Ca, which make up close to 90% of the parent rock, are removed and the less mobile Ni (and Fe) are retained. Through this process the Ni content increases

Table 5.6 Comparison of magmatic and lateritic Ni ores

	Magmatic	Lateritic
Origin	Segregation of Ni sulfides from mafic-ultramafic magmas	Concentration of Ni in soil during long weathering in hot humid climates
Ore minerals	Ni-Cu-Fe sulfides	Ni-bearing phyllosilicates
Ore grade	0.5–5%	1–7%
Locations of major deposits	Canada, Russia, Australia, China	New Caledonia, Indonesian, Cuba, Australia
Economic aspects		
– Mining conditions	Underground (rare open cast) mining of discontinuous deposits in hard rock	Open cast mining of continuous layers of shallow, poorly consolidated soil
– Cost of refining	Relatively low	High (depends on oil price)
– Associated bonus metals	Cu, PGE	Co
– Toxic waste products	Sulfur	Acid or strong alkalis (from refining)

from 0.2% to 0.3% in the peridotite to 1–3% in the ores. Cobalt is also concentrated by this process and constitutes a valuable by-product. The rich garnierite ores of New Caledonia develop in the lower part of the saprolith; the currently exploited iron oxide ores are in the lower part of the pedolith.

Box 5.5 Exploration for Nickel Deposits

Imagine that you are a geologist in a major multinational mineral exploration company and that your boss, the exploration manager, has asked you to plan the future exploration program. The company economists have decided that the price of all base metals will double in the next 10 years and the company has decided to explore for nickel. Your tasks are as follows: (1) to decide what type of deposit to target (magmatic or lateritic); (2) which part of the world to conduct your exploration.

As part of your work you will have to decide whether to conduct "greenfields" exploration (i.e. the search for deposits in parts of the world where no deposits are known), or "brownfields" exploration (the search for new deposits or extensions of deposits in regions known to be mineralized). To decide on the region you will need to consider the geology of possible target regions, their ages and tectonic makeup, the types of ultramafic rocks they contain, and the climate over the past 100 million years. To decide on the type of deposit you will need to consider the pros and cons of each type of deposit, and, because the extraction of Ni from laterites requires so much energy, the future oil price.

Summarize your arguments in a single page (the maximum the exploration manager will want to read).

5.5   Other Lateritic Deposits

The accumulation of gold in residual (eluvial) deposits has been described in Sect. 5.2. When the weathering takes place in hot humid climates, thick lateritic profiles build up and when they develop on rocks with a significant gold content, the weathering can transform protore (gold-bearing material in which the grade is too low to mine) to ore. Good examples of this type of deposit are found in West Africa. The same applies to platinum group elements which are concentrated in lateritic soils over ultramafic rocks in the Carajas region of Brazil.

Another type of surficial deposit forms as "calcrete", the name given to calcified, calcite-rich soils that develop in arid environments. Important uranium deposits (Langer Heinrich in Namibia and Yeelirie in Western Australia) have formed by the precipitation of carbonate in surface waters that leads to the concentration of carnotite, the vivid yellow uranium mineral with the composition $K_2(UO_2)_2(VO_4)_2 \cdot 3H_2O$.

5.6   Supergene Alteration

When sulfide ore bodies are exposed at the surface, the sulfide minerals become oxidized and the ore metals leach downward to become concentrated in a layer near the top of the water table (Fig. 5.10). These layers of "supergene enrichment" contain two to five times more ore metals than the primary ore and they are conveniently located close to the surface where they can be recovered at the start of the mining operation. Developing a mine is a long and complicated procedure that requires a huge investment, often of many millions of dollars. The time from the start of the operation to the recovery and sale of the first metals may be 5–10 years, so the possibility of mining a layer of abnormally rich ore provides welcome financial relief for the company and often makes the difference between a viable and an uneconomic operation.

Figure 5.10 shows the profile through the supergene enrichment zone above a copper sulfide ore body. The uppermost layer, of hydrated iron oxides, is called a gossan. These layers are void of ore metals but they have distinctive textures that signal the presence of buried sulfides and they are commonly used in mineral exploration. Two enriched layers underlie the leached zone. The upper zone of "oxidized" ores contains a wide variety of secondary minerals – carbonates, silicates, sulfates, phosphates – that are commonly well crystallized, brightly coloured, and highly prized by mineral collectors. It is underlain by a zone of sulfide enrichment where the original iron-bearing sulfides, such as chalcopyrite ($CuFeS_2$) are replaced by secondary sulfides that are iron-free or iron poor and have high Cu contents. Examples include chalcocite (Cu_2S), covellite (CuS) and bornite (Cu_5FeS_4).

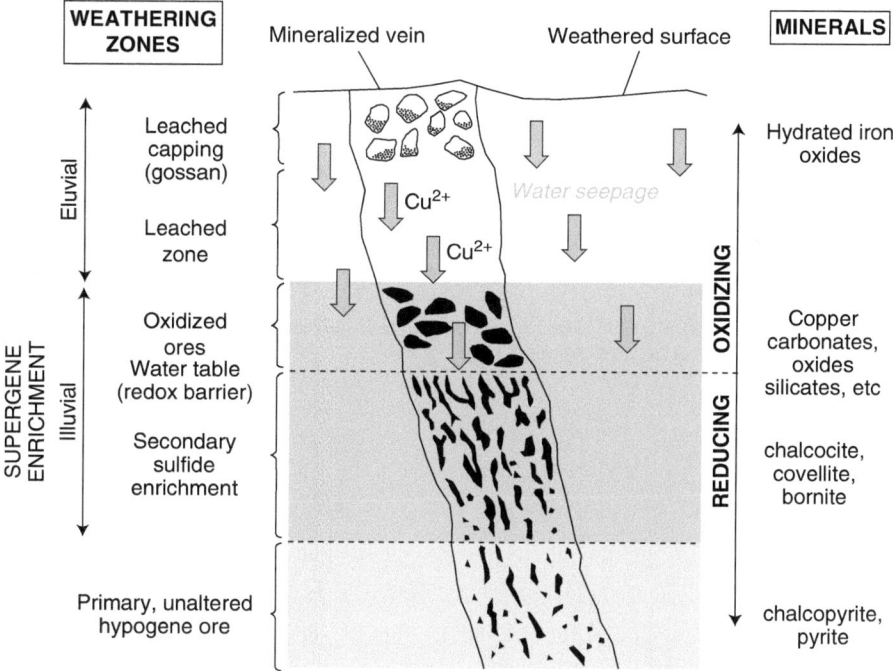

Fig. 5.10 Supergene enrichment zone (From Webb and Rowston (1995))

Perhaps the best-known examples of supergene enrichment zones are those that overlie porphyry copper deposits. Primary ore in the Chuquicamata deposit (discussed in Chap. 4) contains only about 0.8% Cu but it was overlain by a thick layer of supergene enrichment in which the copper grade was 2–3%. In addition, flow of acidic solutions from the main ore body transported dissolved copper downstream where it was redeposited in gravels to form a secondary ore body called the Exotica deposit. This deposit contained an additional 300 t of ore, also with about 2% Cu.

References

Butt CRM, Lintern MJ, Anand RR (2000) Evolution of regoliths and landscapes in deeply weathered terrain – implications for geochemical exploration. Ore Geol Rev 16:167–183

Freyssinet Ph, Butt CRM, Morris RC, Piantone P (2005) Ore-forming processes related to lateritic weathering. Econ Geol 100th Anniversary Volume:681–722

Frimmel HE, Groves DI, Kirk J, Ruiz J, Chesley J, Minter WEL (2005) The formation and preservation of the Witwatersrand goldfields, the largest gold province in the world. In: Hedenquist JW, Thompson JFH, Goldfarb RJ, Richards JP (eds), 100th Anniversary Volume, Society of Economic Geologists, pp 769–797

Garrels RM, Christ CL (1965) Solutions, minerals, and equilibria. Harper & Row, New York, p 450

Garnett, R.H.T., Bassett, N.C., 2005. Placer deposits. Economic Geology 100th Anniversary Volume, pp. 813–843

Klein C, Beukes NK (1993) Sedimentology and geochemistry of glacio-genic late proterozoic iron-formation in Canada. Econ Geol 88:545–565

Lynn MD, Wipplinger PE, Wilson MGC (1998) Diamonds. In: Wilson MGC, Anhaeusser CR (eds), The mineral resources of South Africa. Handbook, Council for Geosciences, 16, 232–258

Schmitz MD, Bowring SA, de Wit MJ, Gartz V (2004) Subduction and terrane collision stabilized the western Kaapvaal craton tectosphere 2.9 billion years ago. Earth Planet Sci Lett 222:363–376

Webb, M. and Rowston, P., 1995. The Geophysics of the Ernest Henry Cu-Au Deposit (NW) Qld. Exploration Geophysics, Vol. 26, pp. 51–59

Chapter 6
The Future of Economic Geology

6.1 Introduction

When we wrote the first edition of this book in 2008–2009, the world was in the depths of the financial crisis and activity of the minerals sector was at a minimum. We nonetheless painted a positive picture of the future of mining, mineral exploration and the study of ore deposits, arguing that the world will always require metals and other mineral products. We recognised that recycling and substitution will meet an increasing proportion of these needs, but the rest must be mined. We talked about whether (not when) our mineral resources will be exhausted and concluded that this is unlikely ever to happen, at least for most metals.

At that time the world was preoccupied with complex issues related to the supply and consumption of petroleum, a product for which it is probable that the rate of production will soon start to decline, if it has not already done so. The rate of this decline and the forces that drive it are subject to major uncertainties. New discoveries of enormous, previously unknown, oil fields off the coast of Brazil and the potential to find other deposits in the Africa and the Arctic suggests that the supply problem may not be as critical as is sometimes made out.

The outlook for the global supply of natural gas has changed more dramatically. Five years ago authorities in the USA were alarmed about the dependence of their country on imported natural gas, often from suppliers in unstable or politically hostile countries. They had started to build a series of new terminals to accommodate tankers that would deliver liquefied natural gas to the USA from the Middle East, Indonesia, Australia and other exporting countries. They estimated that within a decade, the USA would have to import a major portion of its natural gas. At the same time, European leaders were concerned about the dependence of their countries on imports of gas from Russia – concerns exacerbated by the pressure applied by the Russian firm Gazprom on the Ukraine. Then, quite suddenly, technological advances allowed the extraction of gas from a new source – shale gas. This resource had been known for over a century but previously the gas could not be extracted economically from such low-permeability rocks. The new

N. Arndt and C. Ganino, *Metals and Society: an Introduction to Economic Geology*, DOI 10.1007/978-3-642-22996-1_6, © Springer-Verlag Berlin Heidelberg 2012

technology involves horizontal drilling to provide better access to flat-lying sedimentary strata combined with a process, called "fracting" or hydraulic fracturing, in which mixtures of water, sand, and chemical additives are injected at high pressure into the shale to increase its permeability and extract the products – mainly gas and also some oil.

The prices of oil and gas had both dropped drastically during the 2008 financial crisis yet unlike the oil price (and that of most metals), which have since risen dramatically, the gas price has remained at a low level – due to the sudden availability of shale gas. The USA is now expected to satisfy most of its domestic consumption for the next decades, and the potential development of shale gas in Europe, mainly in Poland, Germany, Hungary, Romania and perhaps Great Britain, has loosened the dependence on Russian supplies. There are, to be sure, environmental issues, and ecologists have manifested their concern about the high consumption of water and potential leaks of the chemicals used in the process. But more generally, the global availability of new supplies of natural gas – a clean, low-CO_2 energy source – must be considered a positive development. China, for example, may also possess significant resources of shale gas, and its substitution for the other major domestic energy source – coal – can only be beneficial.

Future decline in the production and consumption of petroleum will probably be driven more by the need to reduce emissions of greenhouse gases than a real exhaustion of resources. The accelerating drive to reduce overall energy consumption will be accompanied by substitution of petroleum by other sources of energy. Nuclear power will play an important role, and a whole array of wind turbines, hydro- or geothermal stations, and solar panels will tap renewable resources. To fuel the nuclear reactors requires uranium; to build the turbines and solar panels, and to improve the performance of cars, electric devices or home heating systems will require a spectrum of hitherto obscure metals. In the following sections we discuss two groups of minerals that have suddenly received global attention and aptly illustrate the challenges of mining in the twenty-first century.

6.2   Rare Earth Elements (REE)

This group of elements, well known to geochemists who use them as tracers of geological processes in the mantle, crust and oceans, is becoming increasingly important in modern industry. They are used in a wide variety of applications, mostly in electronic components, but also in a range of industrial processes. Some typical examples are listed in Table 6.1 and Fig. 6.1 shows where these elements are used in a modern hybrid vehicle.

In geochemical terms, the rare earths are classed as incompatible elements, which means that they become concentrated in the water-rich silicate liquids that remain after a magma has almost completely crystallized or in very low degree partial melts of the mantle. They are present in high abundances in some pegmatites (the products of crystallization of aqueous melts expelled from granitic magmas)

Table 6.1 Uses of the rare earth elements

In 2006, the three main uses for REE in the USA were catalytic converters in cars (25%), catalysts in petroleum refining (22%), and various metallurgical additive and alloys (20%). The emergence of new technologies will rapidly change the situation. Consider, for example, the following list

Lanthanum (La) – water treatment, rechargeable batteries

Cerium (Ce) – glass polishing, heavy 'mud' in oil drilling, catalysers

Neodymium (Nd) – small electric motors, magnets, hard drives in computers, headphones of iPods

Europium (Eu) – red phosphor in flat TV screens

Dysprosium and terbium – alloys and phosphors in lamps and TV tubes, magnets and in the cooling systems of nuclear reactors

A Toyota Prius contains 1 kg of neodymium and 10 kg of lanthanum

A wind turbine contains over 600 kg of rare earths

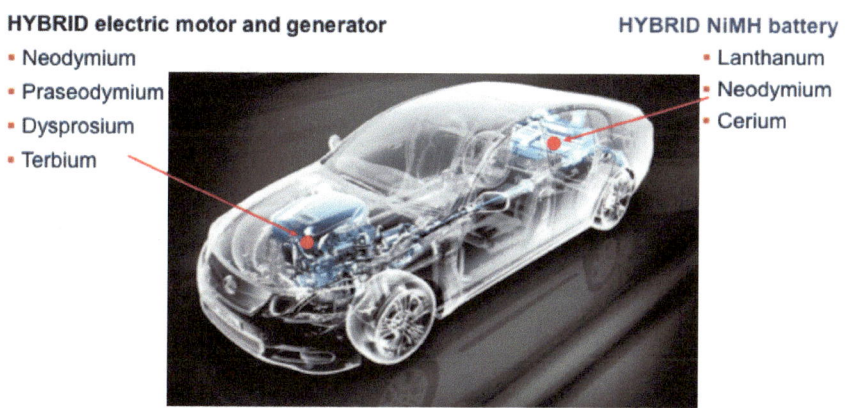

HYBRID electric motor and generator
- Neodymium
- Praseodymium
- Dysprosium
- Terbium

HYBRID NiMH battery
- Lanthanum
- Neodymium
- Cerium

Enabling better emission standards and lower energy consumption

Fig. 6.1 Rare earth elements are essential for the construction of hybrid and electric cars (Reproduced with permission of Matthew James of Lynas corporation)

and in carbonatites and related rocks (alkaline magmas composed of carbonate minerals with few silicates). They also are concentrated in phosphates such as monazite and apatite in detrital sediments, or are absorbed on clays. Table 6.2 lists the REE-bearing minerals in various types of ores.

Bayan Obo, a giant polymetallic (Fe-REE-Nb) deposit of uncertain origin, is located in Inner Mongolia on the northern edge of the North China Craton, about 600 km northwest of Beijing. The deposit was found in 1927 and was first mined for its iron. Current reserves are estimated at about 1.5 billion tons of ore grading 35% Fe. This grade is less than that of the richest iron deposits in Brazil and Australia, but was sufficient for mining in China. In addition to the Fe, the deposit contains a vast amount of REE, a total of some 48 million mt with an average grade of about 6 wt.% of rare-earth oxides. This makes it the world's largest known REE deposit and represents 30–40% of the world's REE resources, depending on the source of information. The deposit also contains large amounts of Nb (Fig. 6.2).

Table 6.2 The growing demand for REE

	Consumption 2010 (%)	Consumption 2015 (%)	Rate of growth (%)
Magnets	21	26	10–15
Alloys	18	19	8–12
Polishing	15	16	8–10
Catalysts	19	15	3–5
Phosphors	7	7	6–10
Glass	9	6	0
Ceramics	6	5	6–8
Other	6	5	6–8

Table 6.3 Compositions of minerals mined for REE

	Minerals rich in light REE			Minerals rich in heavy REE		
	Bastnaesite Bayan Obo (China)	Monazite Orissa (India)	Apatite Arafura (Australia)	Ionic clay Middle Y (China)	Xenotime Pitinga (Brazil)	Fergusonite Thor Lake (Canada)
La	26.9	23.7	20	30	tr	0.3
Ce	50.9	44	48.2	3	0.6	4.4
Pr	5.0	5.85	5.9	7	tr	1.7
Nd	15.2	18.7	21.5	26.4	0.4	15.6
Sm	1.15	5.09	2.4	5.1	0.4	10.4
Eu	0.23		0.41	0.65	tr	1.6
Gd	0.32	1.55	1	4.2	1.1	14.3
Tb	0.03		0.08	0.7	0.8	1.8
Dy	0.09	0.31	0.34	2.9	11.2	9.8
Ho	0.01	tr	tr	0.4	3.4	1.2
Er	0.01	tr	tr	1.4	15.4	1.4
Tm		tr	tr	0.3	3.0	0.7
Yb	0.005	tr	tr	1.1	20.4	4.1
Lu		tr	tr	0.2	2.7	0.7
Y	0.20	0.75		17.3	40.6	29.0

The deposit consists of a series of replacement bodies of disseminated, banded and massive ore, mostly in dolomite marble and more rarely in slate. It is mineralogically very complex and 150 or more minerals have been identified. The principal REE minerals are bastnaesite $(Ce,La,Nd)(CO_3)_2F$, monazite $(Ce,La,Nd)PO_4$ and huanghoite $Ba(Ce,La,Nd)(CO_3)_2F$ (Table 6.2); other ore minerals include those mined for iron (magnetite, hematite, goethite, martite) and for Nb (e.g. columbite $FeNb_2O_2$ and fergusonite $YNbO_4$). Opinion is divided as to the origin of the deposit. There is evidence of transport of the ore metals in hydrothermal fluids and of multiple stages of deposition and replacement of the wall rocks, probably associated with the time of regional metamorphism. Some authors promote a model of carbonate replacement by fluids derived from carbonatitic or felsic alkaline magmas; others consider it a variety of iron-oxide copper gold deposit (Table 6.3).

China contains several other significant REE deposits including Mianning, a carbonatitie-hosted deposit in Sichuan province, and smaller but economically

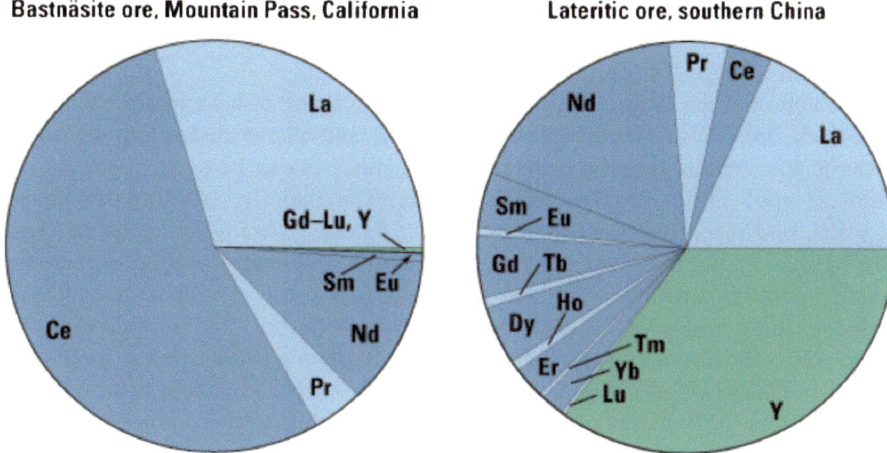

Fig. 6.2 Comparison of the contents of light and heavy REE in two types of ore (From Haxel et al. (2005), http://pubs.usgs.gov/fs/2002/fs087-02/)

significant deposits such as Xunwu and Longnan in Jiangxi province in southeast China. The latter consist of concentrations of ion-adsorption clays that develop in lateritic weathering crusts on granitic and syenitic rocks in the tropical southern part of the country. These oxide ores are economically important because they contain relatively high proportions of the heavy REE (Fig. 6.2). In principle they are easy to mine, being composed of soft materials exposed at the surface; in practice their exploitation, often in an artisanal manner, has generated major pollution and serious environmental problems.

China currently produces most of the world's REE and exerts a major control in global trade in this commodity. During the past decade it has produced REE in large quantities and at low cost, which encouraged the use of these elements for the multitude of applications listed above. More recently, with the explosion of interest in the development of hybrid and electric cars, demand has increased and at the time of writing China had started to exploit its near-monopoly position. It has restricted the supply of the elements in part to incite foreign companies to set up factories within China.

The current crisis in the rare earth elements is due to China's dual control of Bayan Obo and to the ionic clay deposits. The Bayan Obo deposit was opened as an iron mine, and the rare earths were initially produced as a by-product. By benefiting from this situation and using mining practices that would be environmentally unacceptable in other countries, China was able in the early 2000's to undercut the global price and this temporarily drove the major producer – Molycorp's Mountain Pass Mine in the USA – out of business. In 2009 China accounted for 97% of global production, and, by restricting exports, developed a policy to encourage industries requiring the rare earths to relocate their factories to China. Bayan Obo, like many other deposits, produces mainly the light REE (low atomic number), La through to Nd, but the ionic clay deposits contain relatively high concentrations of the heavy rare earths. China is therefore able to satisfy the demand for all types of REE.

Evaluation of the future of REE mining requires that the short, intermediate and long-term prospects are considered separately, and that a distinction be made between the various types of rare earth elements. As can be seen in Table 6.1, different applications require different REE. The majority of currently active deposits, and those likely to come on stream in the next 5 years, produce light REE (Fig. 6.2) and that, with the reopening of the Mountain Pass deposit and the development of new deposits like Mt Weld in Australia, global demand for elements these will soon be satisfied. In contrast, the only major source of the heavy REE (Gd through to Lu) that is currently exploited is the Chinese ionic clay deposits. The newly discovered Kvanefjeld deposits in Greenland contain large amounts of heavy REE, but no realistic estimates see this deposit coming on stream within the next 5–10 years. During this period, there will be a shortfall of these elements and the opportunity for the sole major supplier to control, if not distort, the global market.

Box 6.1 Rocks and Minerals of the Ilimaussaq Intrusion, Host of the Kvanefjeld REE Deposit

There is something about alkaline intrusions that brings out the worst of petrologists and mineralogists. These intrusions contain high abundances of incompatible elements (those elements that become highly concentrated in late-stage silicate liquids) and these elements crystallize as a vast array of obscure minerals. Unlike chemists, who long ago developed a logical and systematic way of naming chemical compounds, mineralogists continue to assign a new name to each newly discovered mineral; and in parallel petrologists assign a new rock name to each assemblage of obscure minerals. The following table lists, for example, a selection of the names assigned to rocks and minerals in the Ilimaussaq Intrusion in Greenland.

Rocks

Naujaite, lujavrite, kakortokite, foyaite as well as more common syenites and granites.

Minerals

Ilimaussaq is the type locality of about 30 minerals. Here is a partial list of 55 of the ca. 200 minerals that have been identified in the intrusion, distinguished because they fluoresce in ultraviolet light.

Albite, analcime, ancylite, apatite, barylite, bertrandite, beryllite, calcite, catapleiite, cerussite, chabazite, chkalovite, elpidite, evenkite, fersmite, fluorite, genthelvite, gmelinite, gonnardite, halloysite, helvite, hemimorphite, leifite, leucophanite, lorenzenite, lovdarite, microcline, montmorillonite, nahcolite, natrolite, natrophosphate, nenadkevichite, pectolite, pectolitemanganoan, polylithionite, senarmontite, sepiolite, sodalite, sorensenite, sphalerite, stilbite, strontianite, terskite, tetranatrolite, thorite, titanite, tugtupite, ussingite, villiaumite, vinogradovite, vitusite-(ce), vuonnemite, whewellite, willemite, zircon.

Only some of these are important in the context of this chapter.

- Steenstrupine is an unusual phospho-silicate mineral that is the dominant host of both REE and uraniumh in the Kvanefjeld deposit
- Cerite and vitusite also host REE in portions of the deposit
- Villuamite (or villiaumite) contain sodium fluoride

The rare earth elements thereby provide a very interesting example of how the development of industry and new technologies require the use of previously little exploited resources, and how the global minerals industry reacts to this demand.

6.3   Lithium

The occurrence and exploitation of this element provide another example of the complications – geological, geographic, economic and political – that will influence the global minerals industry in the first part of the twenty-first century. Until recently this element had been used many specialized applications, but only in relatively small quantities. Some examples are listed in Table 6.4.

Global production of about 20,000 t/year was able to meet this demand over the past decade, but in the near future the situation may change. The push to reduce petrol consumption and CO_2 emissions by the world's growing fleet of automobiles has led to the development of hybrids such Toyota's Prius and a range of fully electric vehicles. Most of these may eventually be equipped with Li-ion batteries, which offer important advantages, including greater power and smaller size and weight, over other types of battery. The battery of a hybrid vehicle contains about 2 kg of Li and that of a fully electric vehicle about 3 kg. To convert the world's fleet of vehicles from petrol to electric would require a vast increase in the demand for Li, up to ten times current production according to some estimates. Where will all this Li come from?

Currently two main types of Li ore are mined. The first consist of deposits of the mineral spodumene, a Li silicate with a pyroxene-like composition ($LialSi_2O_6$), that

Table 6.4 Uses of lithium

– As a flux in aluminium smelting
– As a heat-transfer medium in nuclear reactors (because of its very high specific heat)
– In many types of battery (because of its high electrochemical potential)
– In pharmaceuticals, as mood stabilizer
– As a specialized lubricant
– In alloys with al and Mg to produce strong and light aircraft parts
– In specialized ceramics and glasses (telescope lenses)
– As LiOH which absorbs CO_2 in submarines and spacecraft

Fig. 6.3 Bolivian workers cutting the salt crust at the surface of the Salar de Uyuni, to take samples and prepare for future mining of the deposit. AFP/AIZAR RALDE Le Monde 07/07/2010

occurs in pegmatites; the second is Li carbonate which occurs in evaporitic sediments and in the waters of high-altitude lakes. Past production has been mainly from spodumene, but this has been largely supplanted by the second source, because, just as with Ni ores, the energy requirement to refine the hard silicate mineral, usually in underground mines, is greater than for the alternative. at present about 75% of the world's Li reserves are in South America, in the andean "altiplano", the high flat plain that extends through three countries, Bolivia, Chile and Argentina. Geological factors, such as the presence of siliceous volcanic rocks that are the source of Li, and climatic conditions favour the concentration of Li in the lakes of the altiplano. The high altitude, strong winds and arid climate promote rapid evaporation of the run-off from infrequent storms into closed basins where Li accumulates in lake waters and sediments. Lithium is separated from the brine by a process that starts by allowing the evaporation of the brine in closed pens, very like the extraction of sea salt (Fig. 6.3). The Li is then extracted from the concentrated brine and separated from other salts by a series of chemical reactions. The process is long and drawn out (1–2 years are required for the initial evaporation stage, but relatively cheap.

The concentration in this part of the world of a metal that may become essential for global industry raises numerous questions. More than half the total resource is located in Bolivia, a country with a long, troubled history of mining and mineral exploitation. From the sixteenth to early nineteenth century the Spanish colonialists ruthlessly exploited the incredibly rich silver deposits of "Cerro Rico", shipping most of the wealth back to Spain but briefly making Potosi in the high Andes one of the richest cities in the world. Following the Bolivian revolution and through to the present, the mineral deposits of the country have been managed or mismanaged by a succession of owners. During long periods foreign companies were in charge, and during these periods much of the wealth left the country; and during alternating periods when the mines were nationalized, inefficiency and corruption prevented the local population from receiving much of the wealth generated by the industry. Potosi is now a sad and dilapidated place as all its fine colonial buildings fall into disrepair.

In 2005, Evo Morales was elected the first indigenous president of the country and he immediately took steps to nationalize the oil, gas and mining industry. He has launched an active campaign to renegotiate contracts with the foreign buyers of these natural resources that guarantee that a far greater proportion of the wealth remains in the country.

Several attempts had been made in the past to develop the world's largest lithium deposits, which are found in the Salar de Uyni saltpans in central Bolivia, but each has failed for various political and economic reasons. at the time we wrote this book negotiations were underway to raise the funds needed to develop the deposits and thus to help meet the expected demand for Li batteries, but progress has been slow. Bolivia, a very poor country, does not have the millions of dollars needed to start the operation and foreign sources are reluctant to invest in a country where the political climate, from their point of view, is so uncertain. On one hand the government has said that they will oppose any future program in which cheap Bolivian resources are used to build expensive cars in rich countries; on the other hand the governments of the latter countries do not wish to see Bolivia set up a stranglehold on a energy product that in some ways would be comparable to that of the Middle East oil producers.

Another factor is the composition of the material that is mined. Although the Salar de Uyni deposit contains the greatest tonnage of Li, the ore has a relatively high Mg/Li ratio. Mg is not recovered and has to be deposed of as a waste product. On the other hand the saltpans contain large amounts of potassium salts, which are used as fertiliser, and sodium salts, which could be used in industry, if it could be transported to the places were it would be consumed. at its location high in the sparsely populated Andes, the Salar de Uyni deposits are far from potential consumers.

Environmental issues compound the problem. The lakes of the altiplano constitute a unique ecological system that hosts unique fauna, including large flocks of particularly pink flamingos. The extent to which mining would disrupt these systems is unknown but is likely to be substantial, thus adding an additional reason for the Bolivian government to resist large-scale industrialization of the region. The growing tourism industry in the region also opposes any move to mine the deposits.

Meanwhile, as the situation in Bolivia remains unresolved, Chile and Argentina, which both have governments that are far more open to mining, have developed their segments of the altiplano deposits. In 2010, Chile produced 60% of the world's lithium from its Salar de Atacama deposits in the north of the country. The Greenbushes spodumene deposit in Australia is another important producer and some 70 projects are currently underway to search for or develop deposits in other parts of the world. Major brine resources probably exist in Tibet and Afghanistan

Fig. 6.4 Kryptonite the
mineral that steals
Superman's strength, has the
composition
LiNaSiB$_3$O$_7$(OH), identical
to that of jadarite, a Li
mineral in an ore prospect in
Serbia

and many other types of deposit are known in other areas. Possible sources include
hectorite (a Li-rich clay), geothermal fluids, oilfield brines, and eventually seawater, which contains about 0.17 ppm of Li. at present, the metal cannot be exploited
economically from this source but it is conceivable that future technological
developments will make this possible. Finally Rio Tinto's prospect in the Jadar
Valley of Serbia must be mentioned, not because it is likely to be a major
contributor to the global Li market but because the host mineral jadarite has the
composition LiNaSiB$_3$O$_7$(OH) – identical to that of Superman's kryptonite
(Fig. 6.4).

6.4   Mining and Mineral Exploration in the Future

The graphs reproduced in Chap. 1 starkly illustrate the challenge faced by the
global minerals industry. As world population increases and as people in the third
world aspire to a lifestyle like that in developed countries, the demand for metals
will increase. We have argued that more efficient development of existing deposits,
the opening of new mines and the discovery of new resources will meet this
demand. If the trends that have persisted over the past century continue,
improvements in mining methods and in extraction technology will allow metals
to be extracted from deposits with lower grades than those currently mined, or from
deposits in more hostile or remote locations. The tapping of underwater deposits
such as metal-rich nodules on the seafloor will, sooner or later, provide a vast
additional source of metals such as Ni, Co, Cu, Zn, Mo and Mn, and the mining of

recently formed, still submerged exhalative sulfide deposits will provide a source of Cu, Zn, Pb, au and other metals.

But before these deposits can be mined they must be found. as explained in the first chapter, known reserves of most metals are enough to meet the world's consumption for only the next few decades. At present, and most probably through the first part of this century, national and international mineral exploration companies will conduct the search for new deposits, assisted in many regions by national geological surveys. The goal of most companies will be to find better deposits; i.e. deposits with relatively high grades and geological settings that will allow them to be mined easily and efficiently. The driving force for this search is of course profit, the raison d'être of a private company, but other factors come into play. The mining of a large low-grade ore body involves the extraction of vast amounts of rock, with consequent use of large amounts of energy, water and other resources. To extract copper from ore containing 0.4% Cu produces well over twice as much waste as ore containing 0.8% Cu (over half because the recovery of the metal is not 100% efficient) and the waste must be disposed of or retained. The environmental impact of mining rich ore is therefore less than that of mining poor ore. The environmental consequences of a mining operation now play an important role in the planning and execution of any new mine. One interesting example of these concerns is the developments of processes in which the wastes produced by the mining of deposits in mafic or ultramafic rock are reacted with CO_2 from furnaces or from the air, fixing the greenhouse gas as stable carbonates and thereby offsetting the carbon footprint of the mining operation.

The techniques used in this search for new deposits are rapidly evolving, with ever greater reliance being placed on remote sensing techniques and geophysical methods capable of finding deposits hidden beneath surface layers of sediment, alluvium or deep tropical weathering. The mode of operation of the major companies is currently changing and there has been an unfortunate tendency for them to abandon active exploration and research, leaving these tasks to junior companies and to academics. Yet, at one level or another, geologists will continue to play an important role in the industry.

In the past year, a growing demand for metals had fuelled an increase in metal prices that encouraged companies all around the globe to ramp up their exploration programs. The companies require geologists for this work and they will meet this requirement by hiring competent people wherever they can. One of our reasons for writing this book is to provide at least a basic knowledge of the subject to students graduating from universities. This knowledge should prove useful not only for those few students who find employment in the industry, but also for all the others who, no matter which profession they find themselves in, should know a little about the role of metals in our society and about how the ore that yield them form and are mined.

References

Barnes HL (1979) Geochemistry of hydrothermal ore deposits. Wiley, New York, 997 p

Brimhall GH, Crerar DA (1987) Ore fluids: magmatic to supergene. Rev Miner 17:235–321

Butt CRM, Lintern MJ, Anand RR (2000) Evolution of regoliths and landscapes in deeplyweathered terrain - implications for geochemical exploration. Ore Geol Rev 16:167–183

Cathles LM, Adams JJ (2005) Fluid flow and petroleum and mineral resources in the upper (<20 km) continental crust. Econ Geol 100th anniversary vol, pp 77–110

Chenovoy M, Piboule M (2007) Hydrothermalisme. Spéciation métallique hydrique et systèmes hydrothermaux, Collection grenoble sciences. EDP Sciences, Les Ulis, 624 p

Czamanske GK, Zen'ko TE, Fedorenko VA, Calk LC, Budahn JR, Bullock JH Jr, Fries TL, King BS, Siems DF (1995) Petrography and geochemical characterization of ore-bearing intrusions of the Noril'sk type, Siberia; with discussion of their origin. Res Geol 18:1–48

Dubé B, Gosselin P (2008) Mineral deposits of Canada — greenstone-hosted quartz-carbonate vein deposits. http://gsc.nrcan.gc.ca/mindep/synth_dep/gold/greenstone/index_e.php

Dubé B, Balmer W, Sanborn-Barrie M, Skulski T, Parker J (2000) A preliminary report on amphibolite-facies, disseminated- replacement-style mineralization at the Madsen Gold Mine, Current Research 2000- C17, Geological Survey of Canada, Red Lake, p 12

Ellis AJ (1979) Explored geothermal systems. In: Barnes HL (ed) Geochemistry of hydrothermal ore deposits, 2nd edn. Wiley, New York, pp 632–683

Evans AN (1993) Ore geology and industrial minerals: an introduction. Blackwell Publishing Company, Boston, 390 pp

Franklin JM, Gibson HL, Jonasson IR, Galley AG (2005) Volcanogenic massive sulfide deposits. Econ Geol 100th anniversary vol, pp 523–560

Freyssinet P, Butt CRM, Morris RC, Piantone P (2005) Ore-forming processes related to lateritic weathering. Econ Geol 100th anniversary vol, pp 681–722

Frimmel HE et al (2005) Archean atmospheric evolution: evidence from the Witwatersrand gold fields, South Africa. Earth-Sci Rev 70:1–46

Galley AG, Hannington MD, Jonasson IR (2007) Volcanogenic massive sulphide deposits. In: Goodfellow WD (ed) Mineral deposits of Canada: geological association of Canada, vol 5. Geological Association of Canada, St. John's, pp 141–161

Garrels RM, Christ CL (1965) Solutions, minerals, and equilibria. Harper & Row, New York, p 450

Goldhaber MB, Reynolds RL, Rye RO (1978) Origin of a south Texas roll-type uranium deposit: II, sulfide petrology and sulfur isotope studies. Econ Geol 73:1690–3705

Goodfellow WD, Lydon JW (2007) Sedimentary exhalative (SEDEX) deposits. In: Goodfellow WD (ed) Mineral deposits of Canada: geological association of Canada, vol 5. Geological Association of Canada, St. John's, pp 163–183

N. Arndt and C. Ganino, *Metals and Society: an Introduction to Economic Geology*,
DOI 10.1007/978-3-642-22996-1, © Springer-Verlag Berlin Heidelberg 2012

Hannington MD, Galley AG, Herzig PM, Petersen S (1998) Comparison of the TAG mound and stockwork complex with Cyprustype massive sulfide deposits; proceedings of the ocean drilling program, scientific results, vol 158. College Station, pp 389–415

Hedenquist JW, Henley RW (1985) Hydrothermal eruptions in the Waiotapu geothermal system, New Zealand: origin, breccia deposits and effect of precious metal mineralization. Econ Geol 80:1640–1666

Herrington R, Maslennikov V, Zaykov V, Seravkin I, Kosarev A, Buschmann B, Oregeval JJ, Holland N, Tesalina S, Nimis P, Armstrong R (2005) Classification of VMS deposits: lessons from the South Uralides. Ore Geol Rev 27(1–4):203–237. doi:10.1016/j.oregeorev.2005.07.014

Irvine TN (1980) Magmatic infiltration metasomatism, double-diffusive fractional crystallization, and adcumulus growth in the Muskox intrusion and other layered intrusions. In: Hargreaves RB (ed) Physics of magmatic processes. Princeton University Press, Princeton, pp 325–383

Irvine TN (1977) Origin of chromitite layers in the Muskox intrusion and other stratiform intrusions: a new interprétation. Geology 5:273–277

Jefferson CW, Thomas DJ, Gandhi SS, Ramaekers P, Delaney G, Brisbin D, Cutts C, Quirts D, Portella P, Olson RA (2008) Mineral deposits of Canada – unconformity associated uranium deposits. http://gsc.nrcan.gc.ca/mindep/synth_dep/uranium/index_e.php

Kirkham RV, Sinclair WD (1988) Comb quartz layers in felsic intrusions and their relationship to the origin of porphyry deposits. In: Taylor RP, Strong DF (eds) Recent advances in the geology of granite-related mineral deposits, vol 39, The Canadian Institute of mining and metallurgy. Canadian Institute of mining and metallurgy, Montréal, pp 50–71

Klein C, Beukes NK (1993) Sedimentology and geochemistry of glaciogenic late proterozoic iron-formation in Canada. Econ Geol 88:545–565

Krupp RE, Seward TM (1987) The Rotokawa geothermal system, New Zealand: an active epithermal gold-depositing environment. Econ Geol 82:1109–1129

Leach D, Sangster D, Kelley K, Large RR, Garven G, Allen C, Gutzmer J, Walters S (2005) Sediment-hosted lead-zinc deposits: a global perspective. Econ Geol 100th anniversary, pp 561–607

Little CTS, Magalashvili AG, Banks DA (2007) Neotethyan late cretaceous volcanic arc hydro-thermal vent fauna. Geology 35:835–838

Lowell JD, Gilbert JM (1970) Lateral and vertical alteration-mineralization zoning in porphyry ore deposits. Econ Geol 65:373–408

Lynn MD, Wipplinger PE, Wilson MGC (1998) Diamonds. In: Wilson MGC, Anhaeusser CR (eds) The mineral resources of South Africa, vol 16, Handbook, council for geosciences. Council for Geosciences, Silverton, pp 232–258

Ossandon CG, Freraut RC, Gustafson LB, Lindsay DD, Zentilli M (2001) Geology of the Chuquicamata mine: a progress report. Econ Geol 96(2):249–270. doi:10.2113/96.2.249 DOI:dx.doi.org

Paradis S, Nelson JL (2007) Metallogeny of the Robb Lake carbonate-hosted zinc-lead district, northeastern British Columbia. In: Goodfellow WD (ed) Mineral deposits of Canada, vol 5, Geological association of Canada. Geological Association of Canada, St. John's, pp 633–654

Poulsen KH, Ames DE, Lau MHS, Brisbin DI (1986) Preliminary report on the structural setting of gold in the Rice Lake area, Uch subprovince, Current Research 1986, Geological Survey of Canada, Southeastern Manitoba, pp 213–221

Ripley EM, Lightfoot PC, Li C, Elswick ER (2003) Sulfur isotopic studies of continental flood basalts in the Noril'sk region: Implications for the association between lavas and ore-bearing intrusions. Geochimica et Cosmochimica Acta 67:2805–2817

Robb LJ (2005) Introduction to ore-forming processes (2005). Blackwell Publishing, Malden, 373 pp

Schmitz MD, Bowring SA, de Wit MJ, Gartz V (2004) Subduction and terrane collision stabilized the western Kaapvaal craton tectosphere 2.9 billion years ago. Earth Planet Sci Lett 222:363–376

Sillitoe RH (2010) Porphyry copper systems. Econ Geol 105:3–41

Simmons SF, Browne PR (2000) Hydrothermal minerals and precious metals in the Broadlands-Ohaaki geothermal system: implications for understanding low-sulfidation epithermal environments. Econ Geol 95:971–999

Sinclair WD (2007) Porphyry deposits. In: Goodfellow WD (ed) Mineral deposits of Canada: a synthesis of major deposit-types, district metallogeny, the evolution of geological provinces, and exploration methods, vol 5, Geological association of Canada, mineral deposits division. Geological Association of Canada, St. John's, pp 223–243

Von Damm KL (1990) Seafloor hydrothermal activity: black smoker chemistry and chimney. Ann Rev Earth Planet Sci 18:173–204

Webb M, Rowston P (1995) The Geophysics of the Ernest Henry Cu-Au Deposit (NW) Qld. Explor Geophys 26(2/3):51–59

Index